£1.80

Photomicrographs of the flowering plant

D1246131

£1.80

Photomicrographs of the flowering plant

Photomicrographs of the flowering plant

A. C. Shaw
Head of the Biology Department, The Skinners' School, Tunbridge Wells

S. K. Lazell, A.R.P.S.
Medical Photographer, Tunbridge Wells Hospital Group

G. N. Foster
Junior Research Associate, School of Agriculture,
University of Newcastle-upon-Tyne

LONGMAN

Longman Group Ltd
London
Associated companies, branches and representatives throughout the world

First published 1965
Eighth impression 1977

ISBN 0 582 32251 0

Printed in Hong Kong by
Sheck Wah Tong Printing Press Ltd

Contents

Preface

The aim of this book is to present to students at Advanced Level Botany and Biology a set of photomicrographs on plant anatomy side by side with labelled drawings from which they can interpret what they see under the microscope. This study of typical plant material should make it easy for students to identify unknown material when they come to cut it.

We wished to make each drawing an exact replica of the photograph and for this reason we obtained the drawing by using a faint print, drawing on this with Indian ink and then bleaching out the photograph. Guide lines have been drawn in red, not because we expect the student to use red, but for the sake of clarity on the printed page.

The traditional way to examine a plant section is to view it under the high power to identify the tissues and then to draw a low power plan of their distribution. This is followed by a high power drawing of a sector of the slide, in which cells are drawn from each tissue in the sector, first the middle lamellae and then the cell walls. We have not been able to keep rigidly to this system for reasons of book production. For example, full sectors would be too large, cell walls are sometimes so small that we have only given the middle lamellae, and we have not given the high power of a low power where a plan is sufficient.

These photographs and drawings are obviously not a substitute for personal work. Our experience has shown, however, that the use of labelled photomicrographs has speeded up the student's drawing, and enabled him within the time of his Advanced Level course to carry out more investigation of unknown material than is usual.

Our slides are ones which are easily purchased, or easily cut in the laboratory. The slides we purchased came from Messrs. Flatters and Garnett Ltd., or from Messrs T. Gerrard & Co. Ltd. We used for most of our photography a Beck 35 mm eyepiece camera with Ilford Micro-Neg. film. To these suppliers we make grateful acknowledgment and also to the many people who have advised us and criticised this work during its preparation.

A. C. Shaw
S. K. Lazell
G. N. Foster

Tunbridge Wells, 1964.

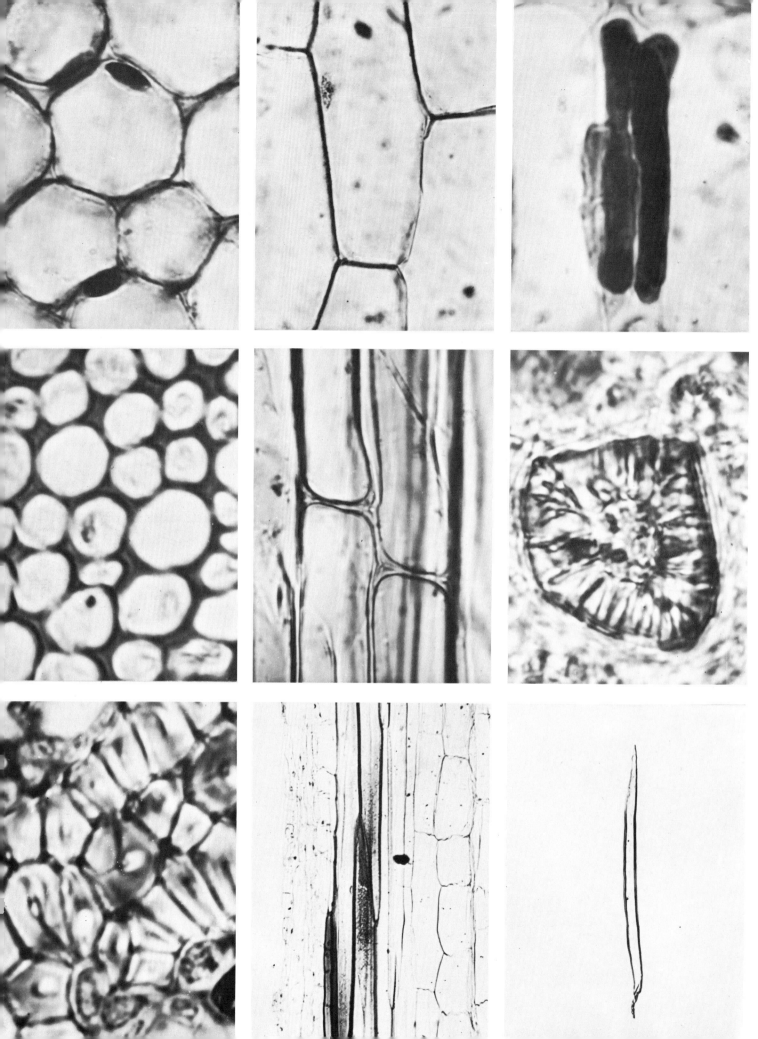

Fig. 1. High power drawings to show the histology of parenchyma, collenchyma and sclerenchyma

nucleus

cytoplasmic lining

Parenchyma in T.S.—*Vicia*

nucleus

cytoplasm

Parenchyma in L.S.—*Vicia*

lumen

lignified cell wall

Sclereids (Stone cells) in L.S.—*Hakea*

angular thickening

protoplast

middle lamella

Collenchyma in T.S.—*Lamium*

middle lamella

angular thickening

Collenchyma in L.S.—*Vicia*

branching simple pits

Sclereid (Stone cell)—*Pyrus*

thick cell wall

lumen

simple pit

Fibres in T.S.—*Tilia*

tapering overlapping end of fibre

Fibres in L.S.—*Zea*

simple pit

thick cell wall

Fibre from maceration of *Tilia* stem

9

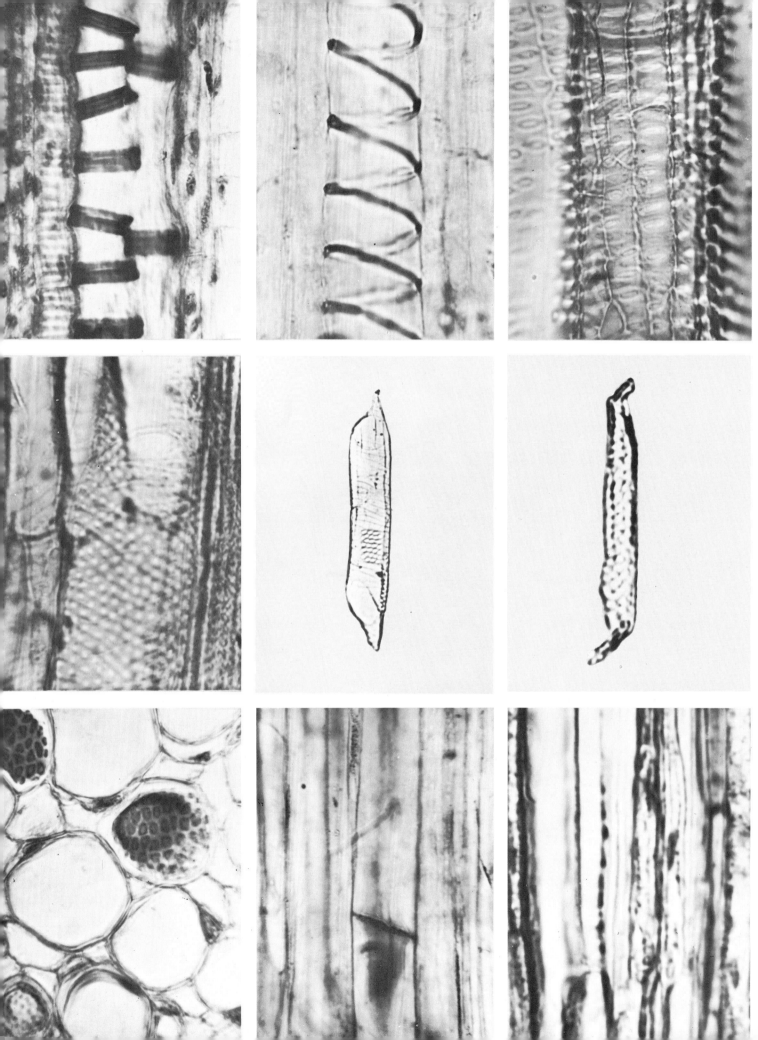

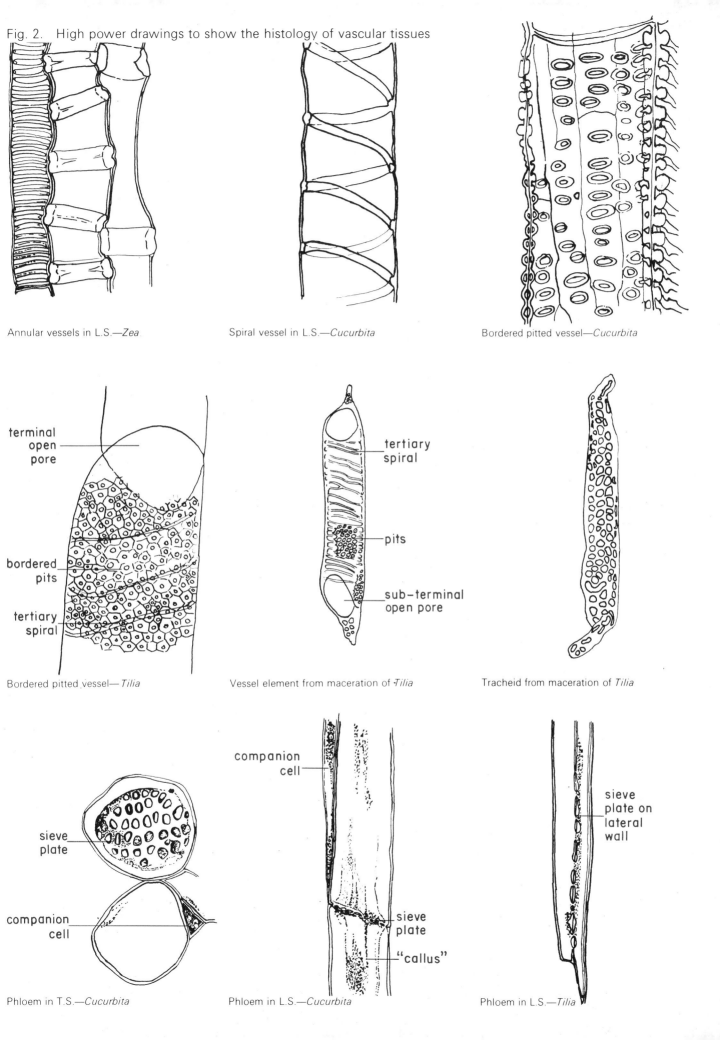

Fig. 2. High power drawings to show the histology of vascular tissues

Annular vessels in L.S.—*Zea*.

Spiral vessel in L.S.—*Cucurbita*

Bordered pitted vessel—*Cucurbita*

terminal open pore

bordered pits

tertiary spiral

Bordered pitted vessel—*Tilia*

tertiary spiral

pits

sub-terminal open pore

Vessel element from maceration of *Tilia*

Tracheid from maceration of *Tilia*

sieve plate

companion cell

Phloem in T.S.—*Cucurbita*

companion cell

sieve plate

"callus"

Phloem in L.S.—*Cucurbita*

sieve plate on lateral wall

Phloem in L.S.—*Tilia*

11

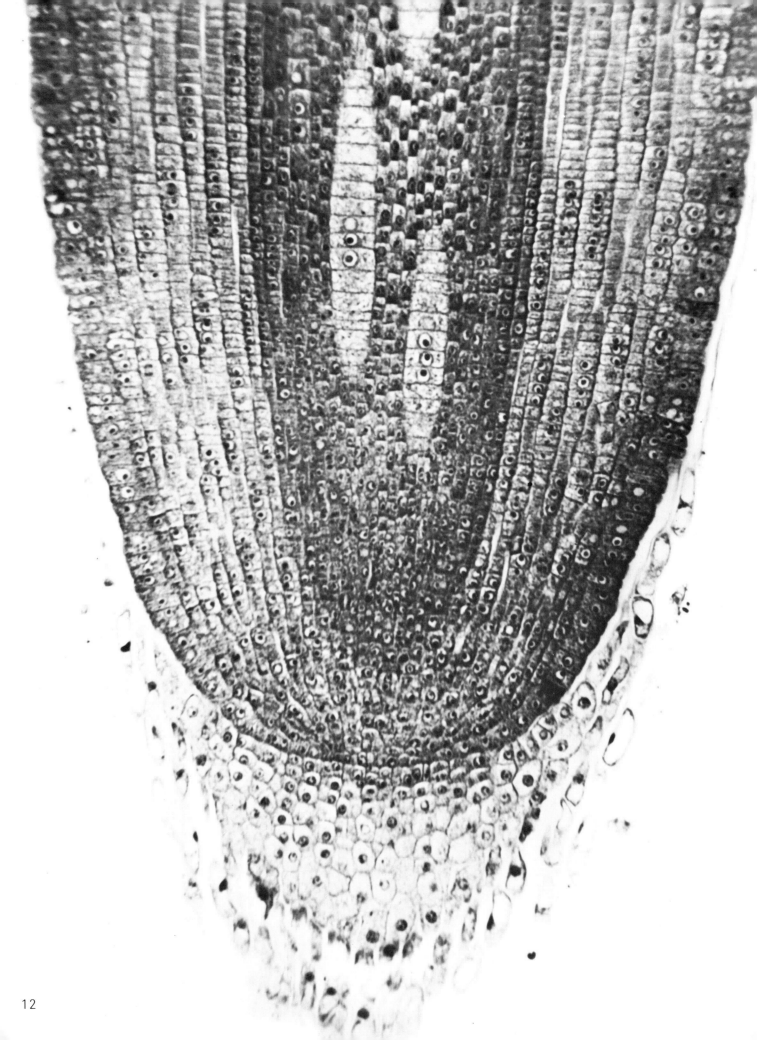

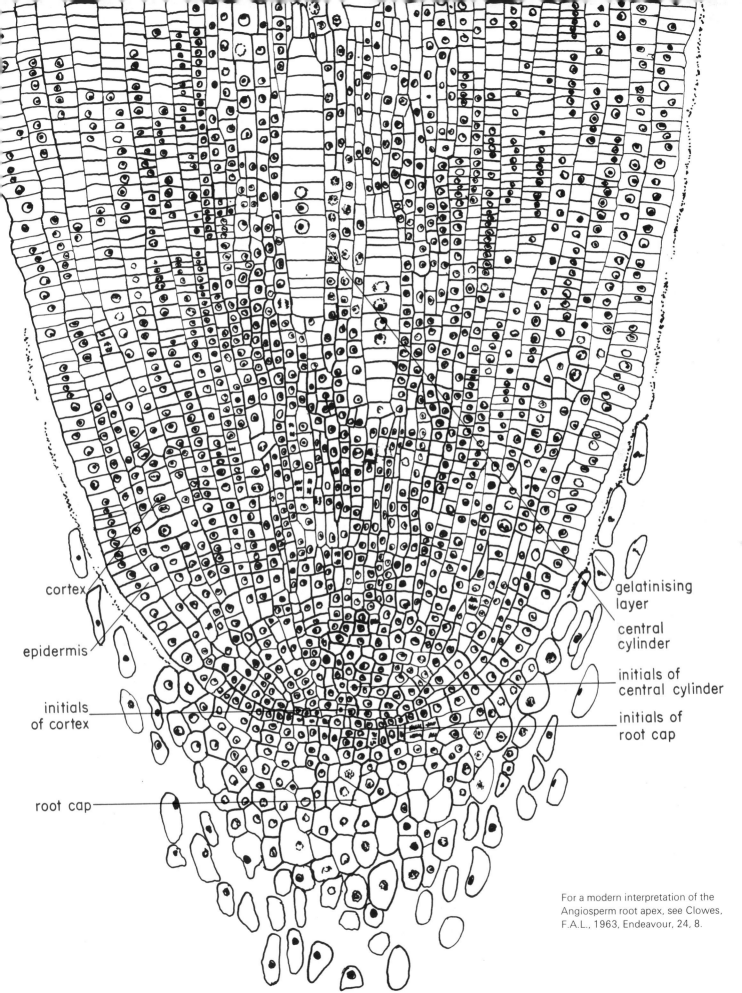

cortex

epidermis

initials
of cortex

root cap

gelatinising
layer

central
cylinder

initials of
central cylinder

initials of
root cap

For a modern interpretation of the
Angiosperm root apex, see Clowes,
F.A.L., 1963, Endeavour, 24, 8.

Fig. 3.   Low power diagram of a longitudinal section through the root apex of *Hordeum murinum*   13

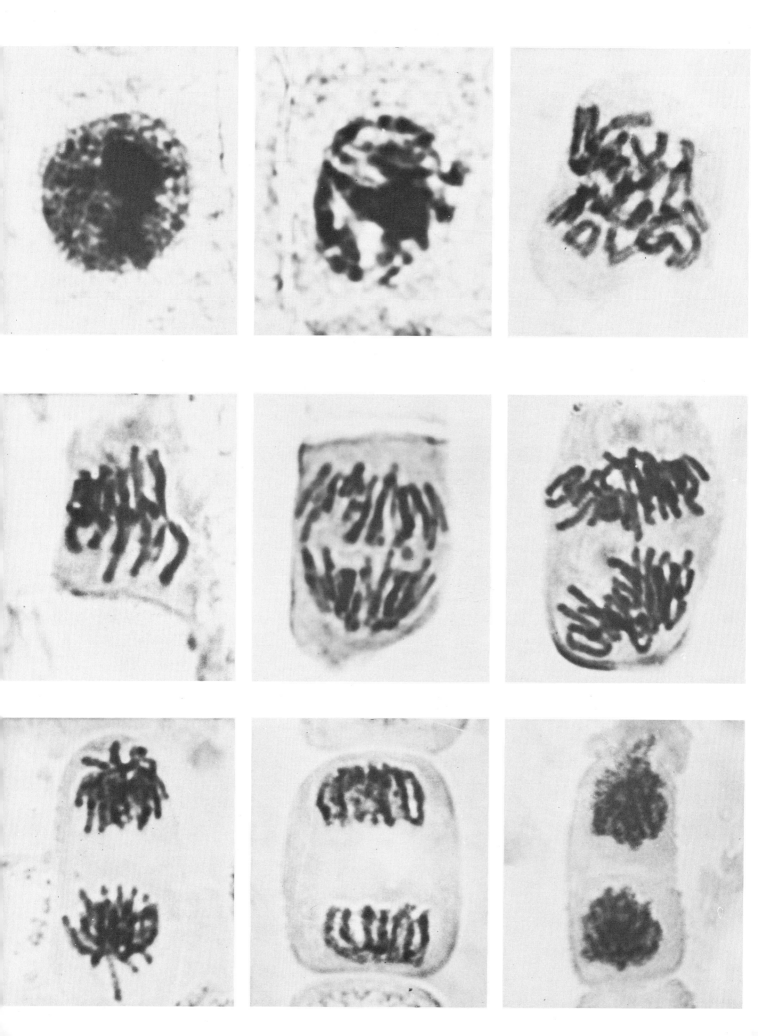

Fig. 4. High power drawings to show stages in mitosis from *Allium* root.
Interphase and prophase from a longitudinal section, other stages from a squash preparation

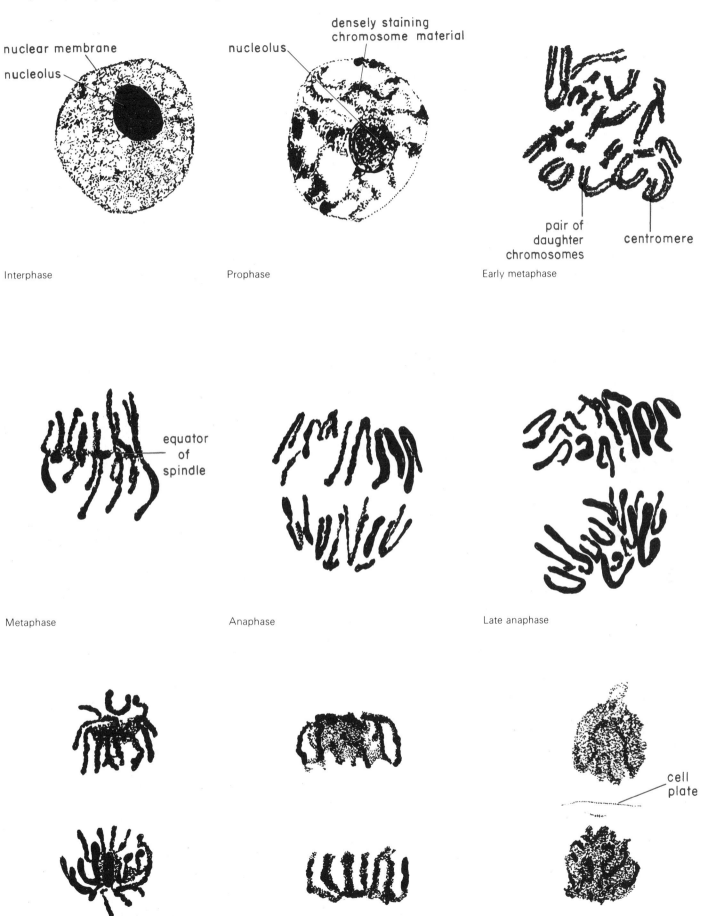

nuclear membrane

nucleolus

Interphase

nucleolus

densely staining chromosome material

Prophase

pair of daughter chromosomes

centromere

Early metaphase

equator of spindle

Metaphase

Anaphase

Late anaphase

Early telophase

Telophase

cell plate

Formation of cell plate

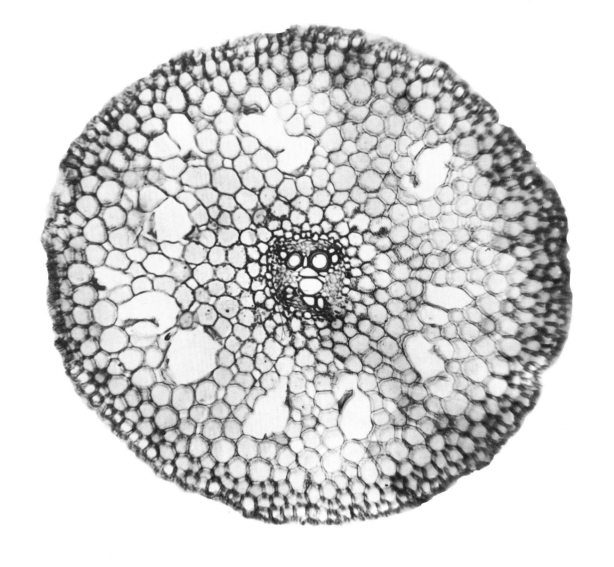

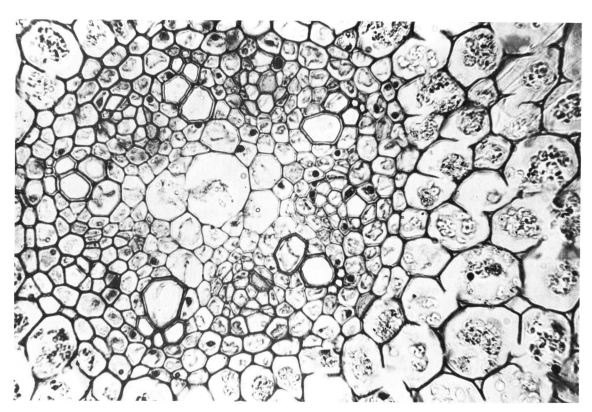

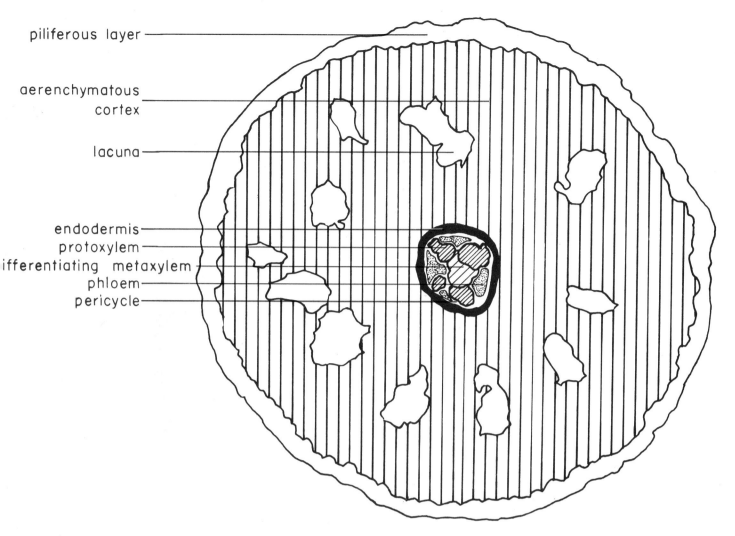

piliferous layer

aerenchymatous cortex

lacuna

endodermis
protoxylem
ifferentiating metaxylem
phloem
pericycle

Fig. 5a.   Low power diagram of a transverse section through a young root of *Ranunculus repens* (tetrarch condition)

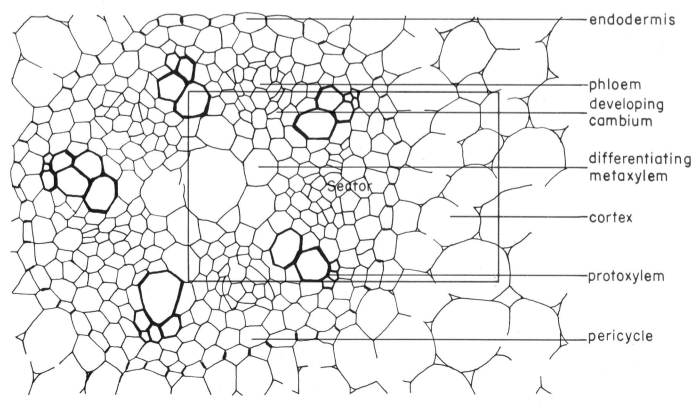

endodermis

phloem
developing cambium

differentiating metaxylem

cortex

protoxylem

pericycle

Sector

Fig. 5b.   Diagram to show further detail of the stele of *Ranunculus repens* (pentarch condition)

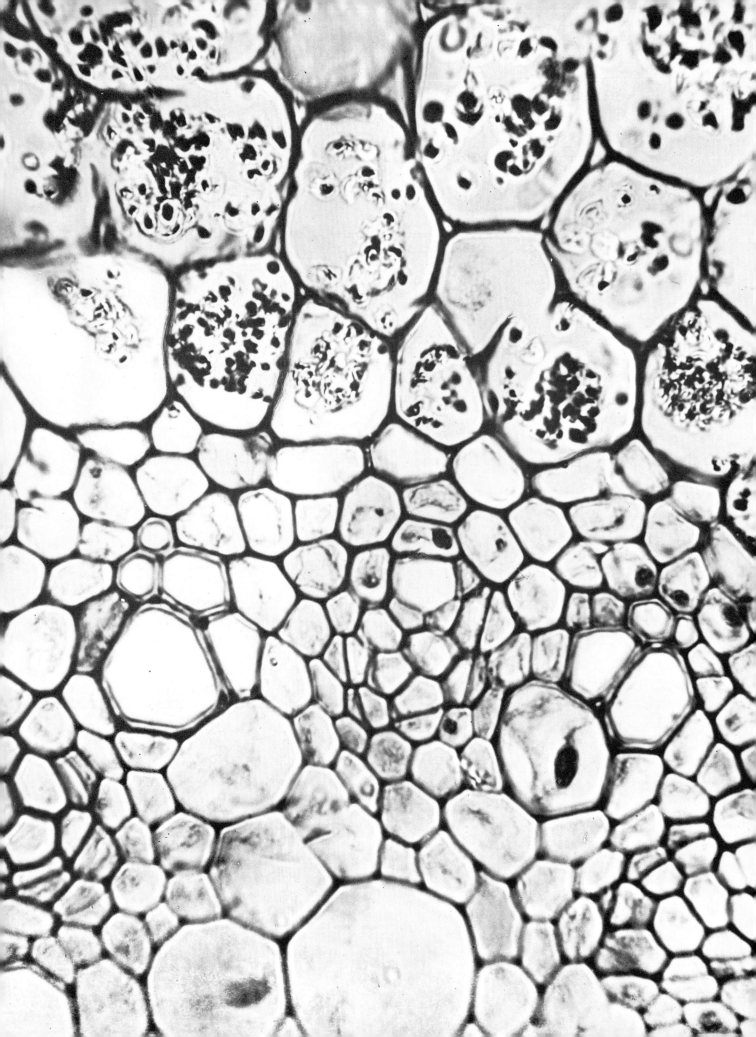

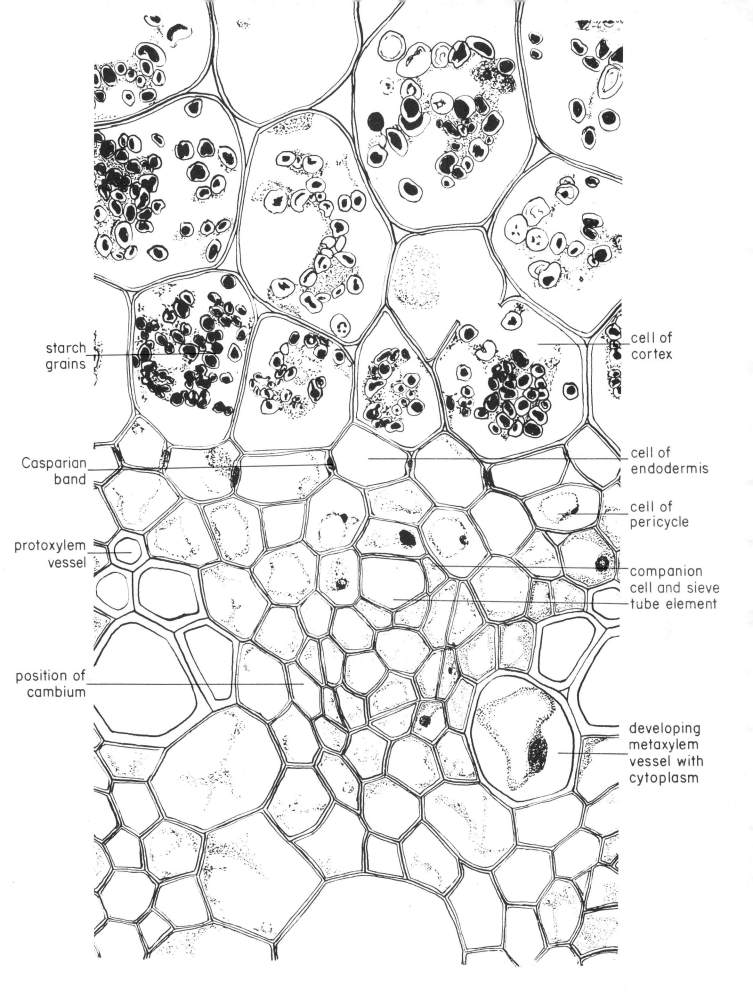

starch
grains

Casparian
band

protoxylem
vessel

position of
cambium

cell of
cortex

cell of
endodermis

cell of
pericycle

companion
cell and sieve
tube element

developing
metaxylem
vessel with
cytoplasm

Fig. 6.   High power drawing of a sector from a transverse section of a young root of *Ranunculus repens*   19

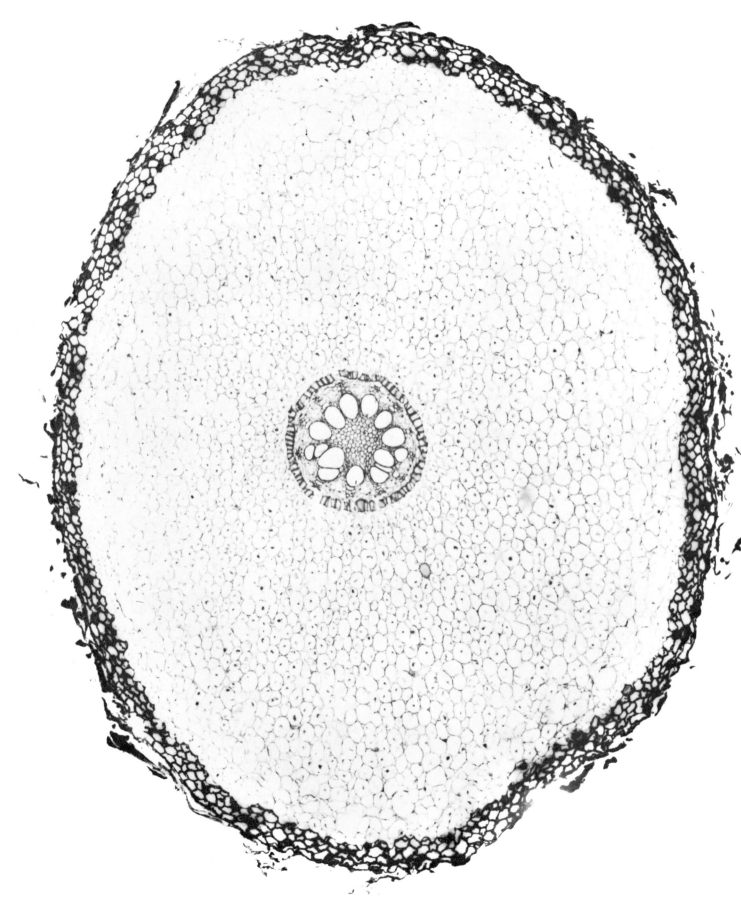

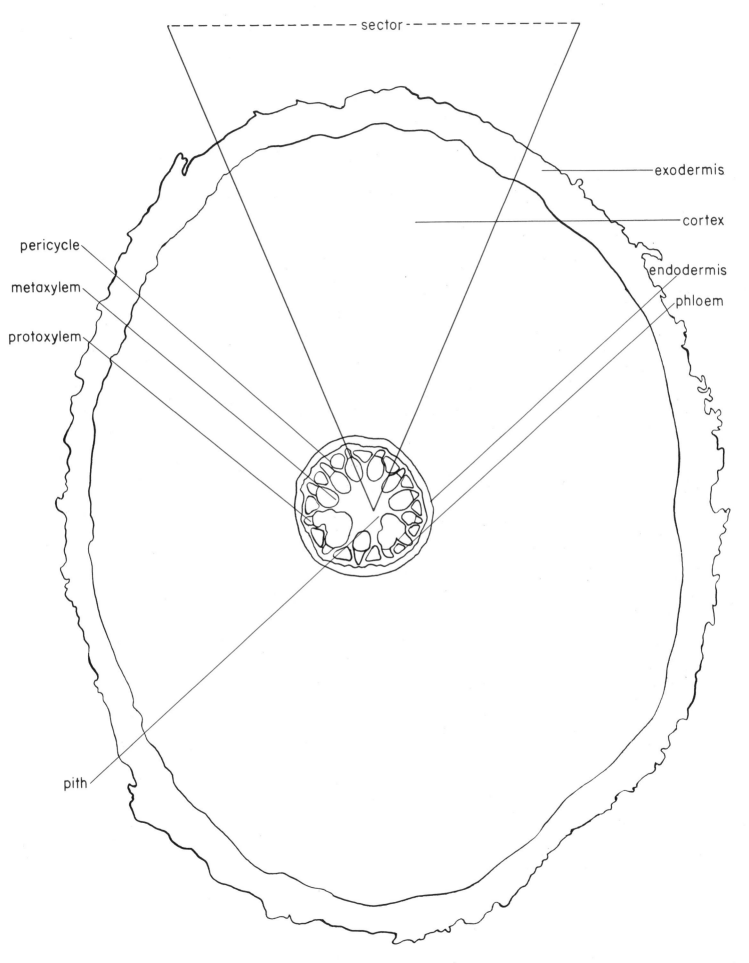

Fig. 7. Low power plan of a transverse section through a root of *Iris germanica*

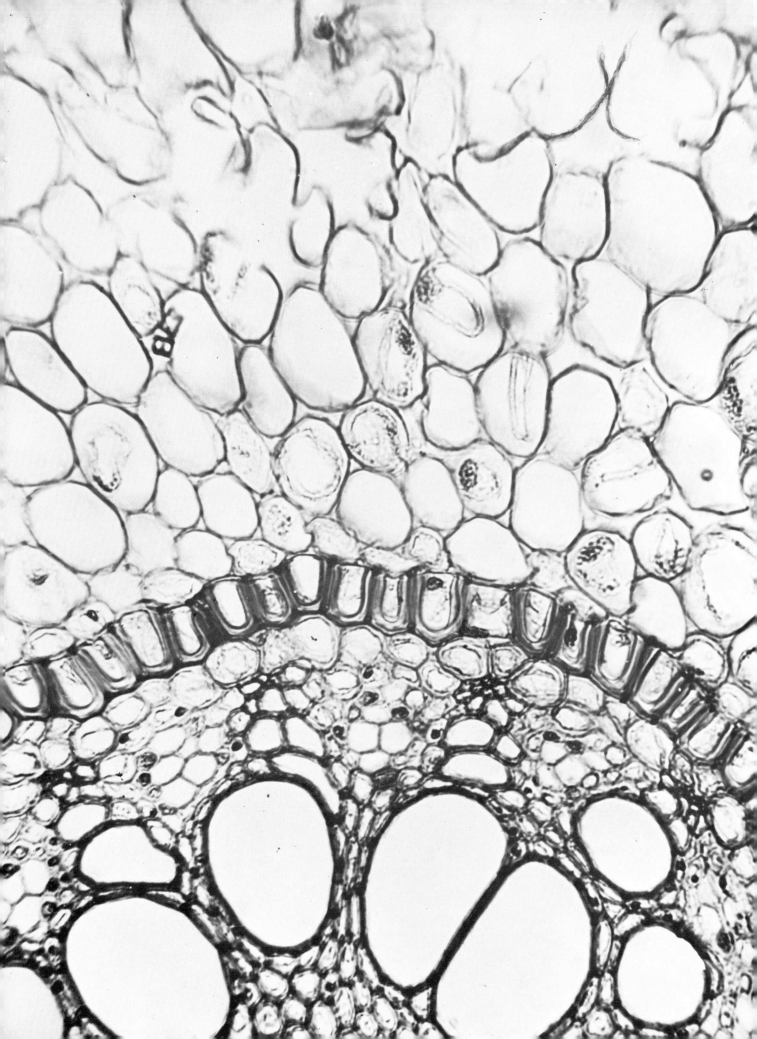

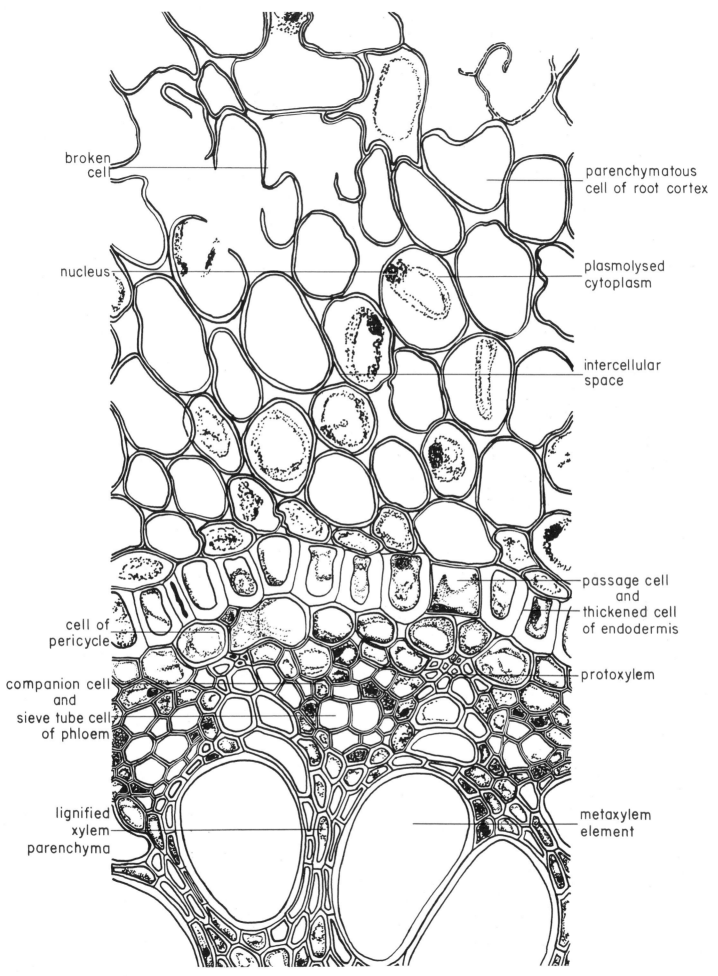

broken cell

parenchymatous cell of root cortex

nucleus

plasmolysed cytoplasm

intercellular space

passage cell and thickened cell of endodermis

cell of pericycle

protoxylem

companion cell and sieve tube cell of phloem

lignified xylem parenchyma

metaxylem element

Fig. 8.   High power drawing of a sector from a transverse section through a root of *Iris germanica*

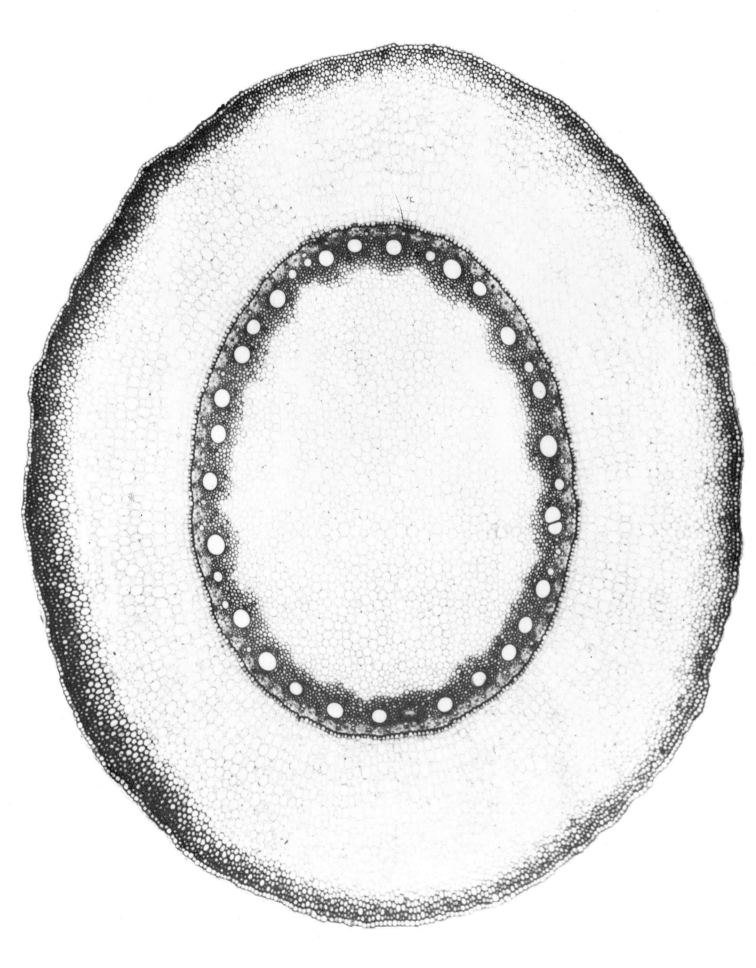

24

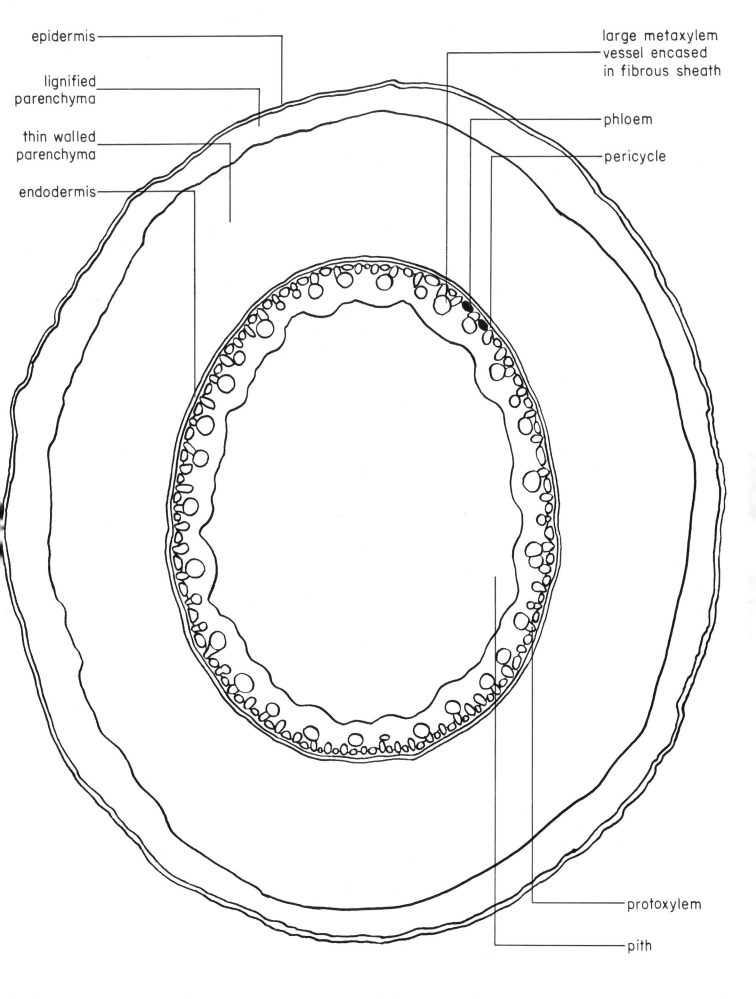

epidermis

lignified
parenchyma

thin walled
parenchyma

endodermis

large metaxylem
vessel encased
in fibrous sheath

phloem

pericycle

protoxylem

pith

Fig. 9. Low power diagram of a transverse section through a buttress root of *Zea mais*   25

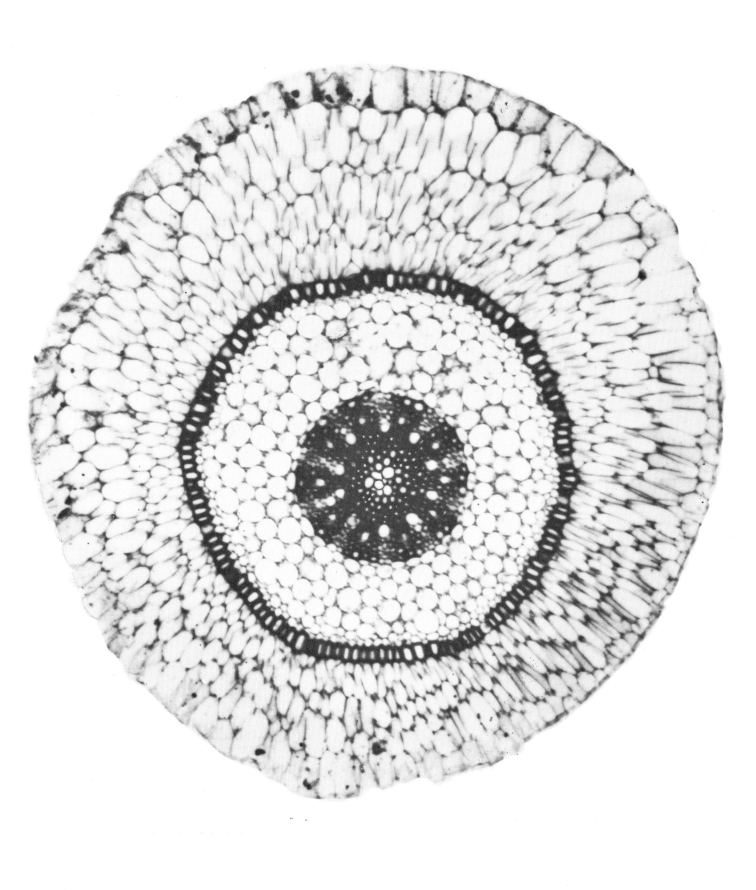

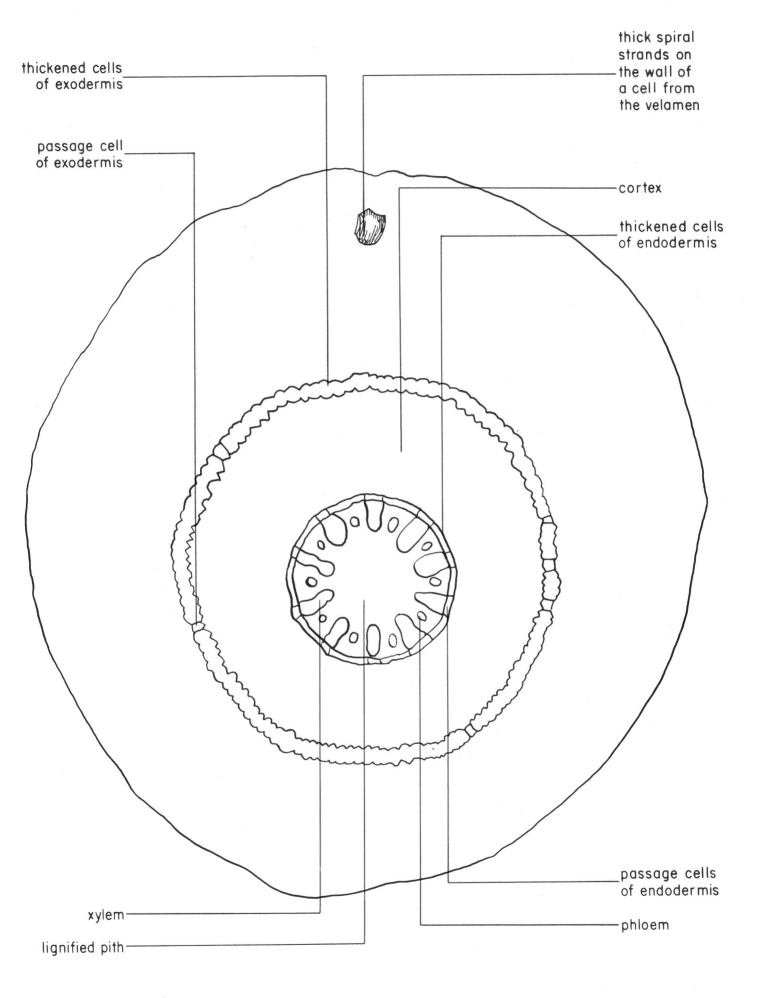

thickened cells
of exodermis

passage cell
of exodermis

thick spiral
strands on
the wall of
a cell from
the velamen

cortex

thickened cells
of endodermis

passage cells
of endodermis

phloem

xylem

lignified pith

Fig. 10. Low power diagram of a transverse section through a "breathing" root of *Dendrobium*, an epiphytic orchid 27

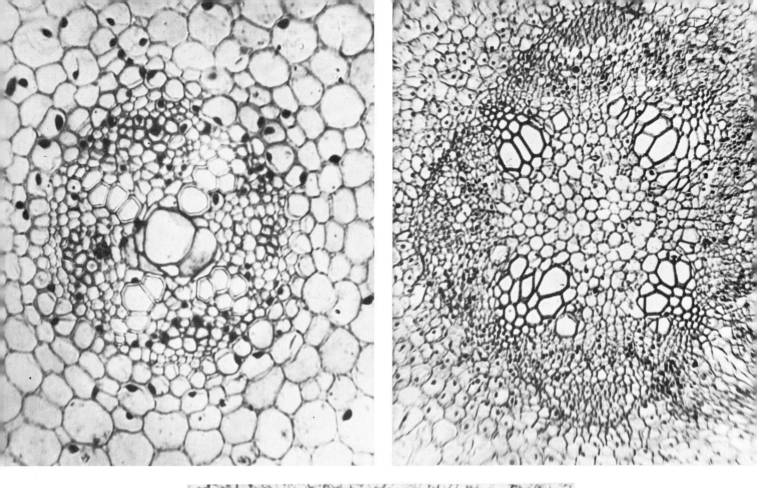
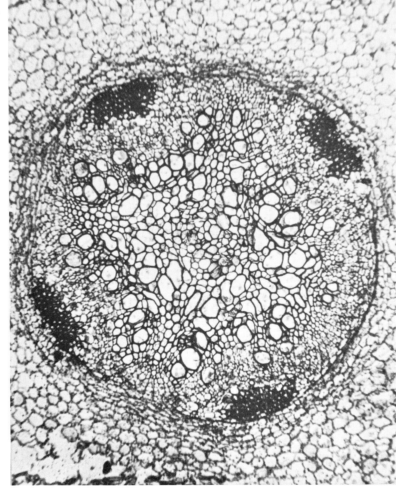

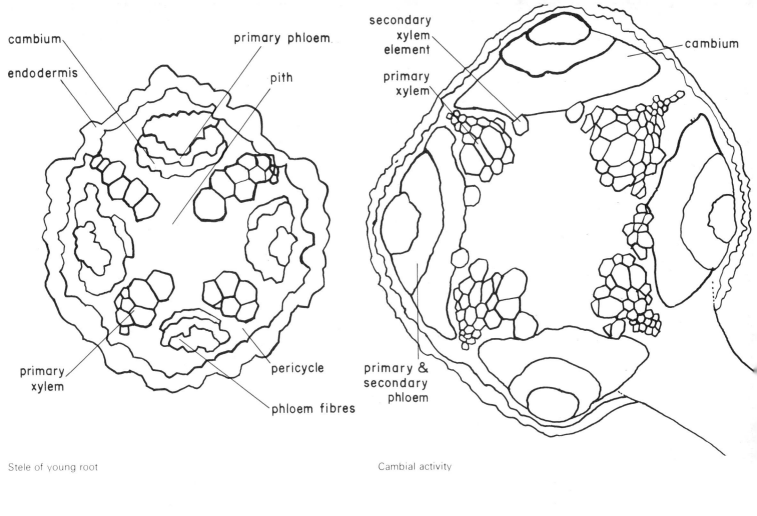

Stele of young root

Cambial activity

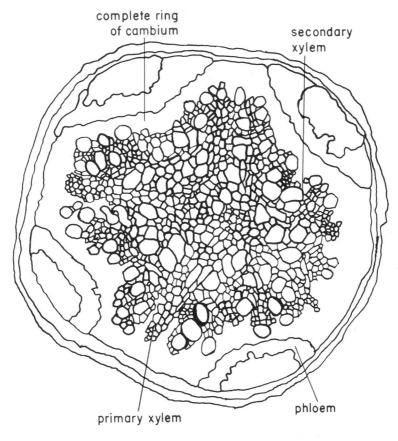

Old root

Fig. 11.   Low power diagrams to show secondary thickening in transverse sections of the root of *Vicia faba*

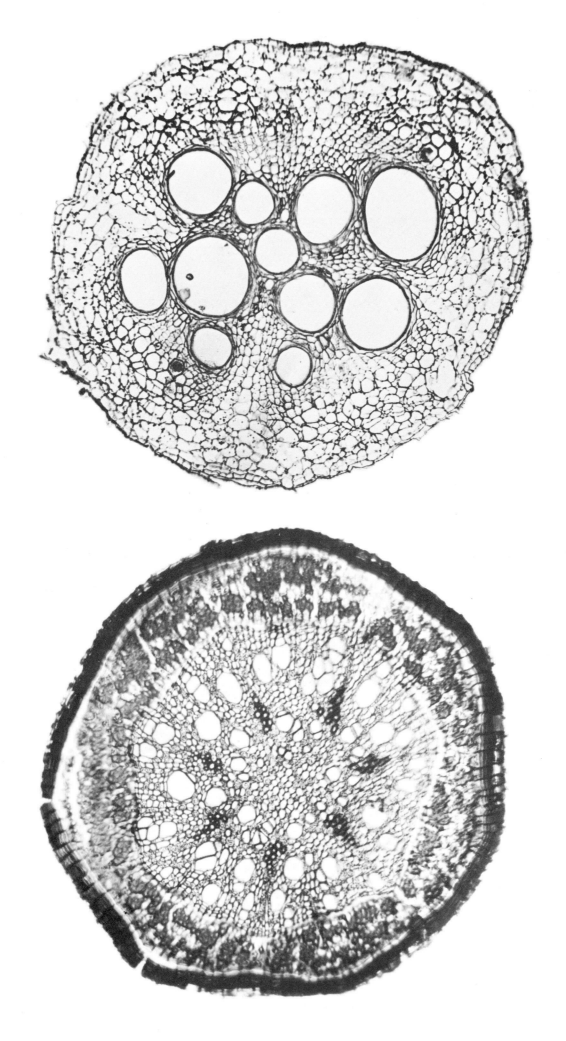

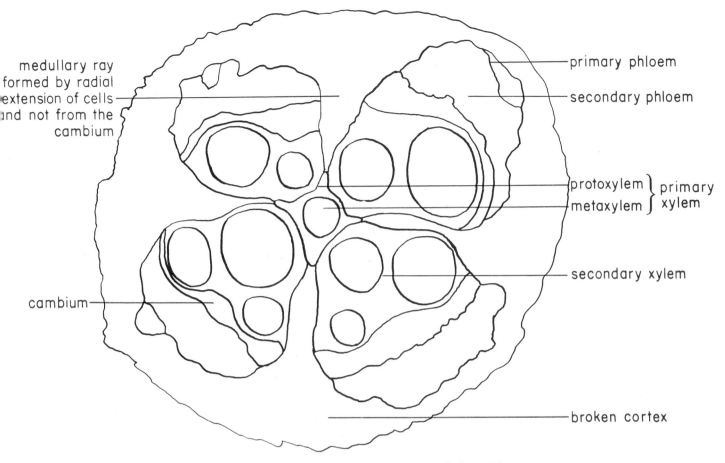

medullary ray formed by radial extension of cells and not from the cambium

primary phloem

secondary phloem

protoxylem } primary
metaxylem } xylem

secondary xylem

cambium

broken cortex

Fig. 12.   Low power diagram of a transverse section through an old root of *Cucurbita pepo*

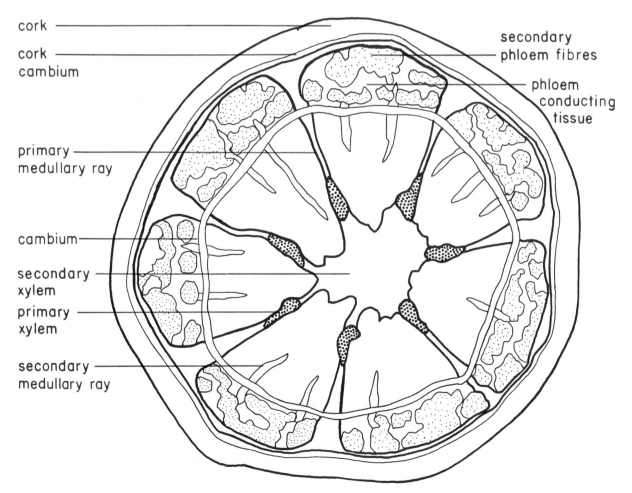

cork

cork cambium

secondary phloem fibres

phloem conducting tissue

primary medullary ray

cambium

secondary xylem

primary xylem

secondary medullary ray

Fig. 13.   Low power diagram of a transverse section through an old root of *Tilia europaea*

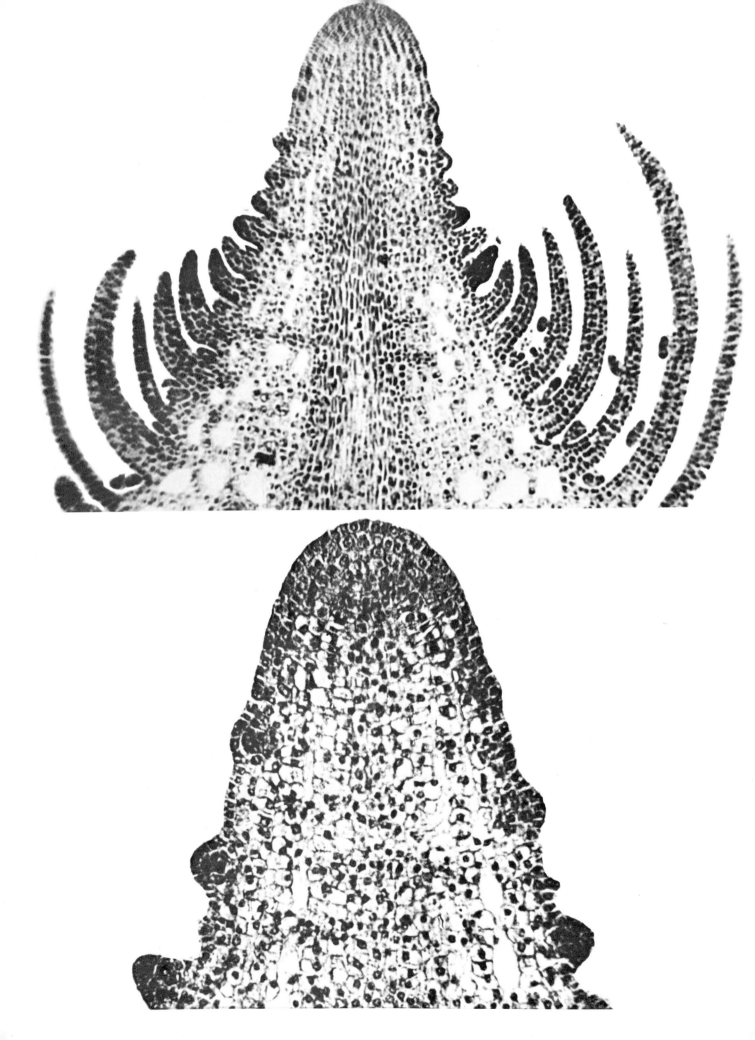

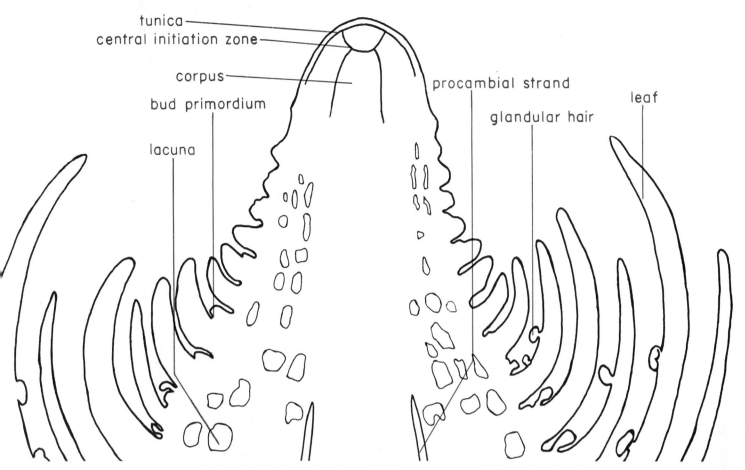

Fig. 14a. Low power diagram of a longitudinal section through the stem apex of *Hippuris vulgaris*

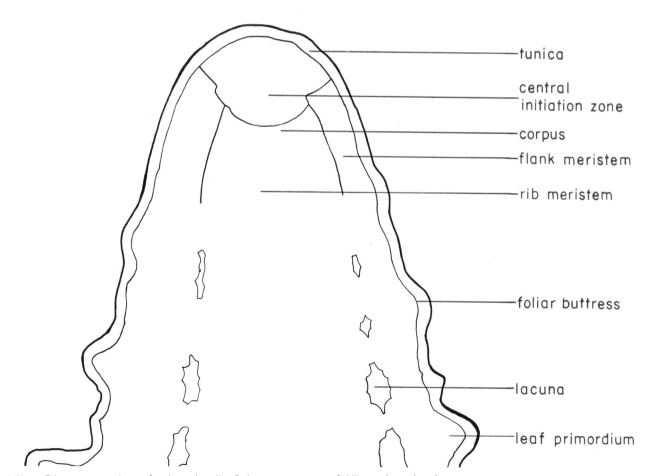

Fig. 14b. Diagram to show further detail of the stem apex of *Hippuris vulgaris*

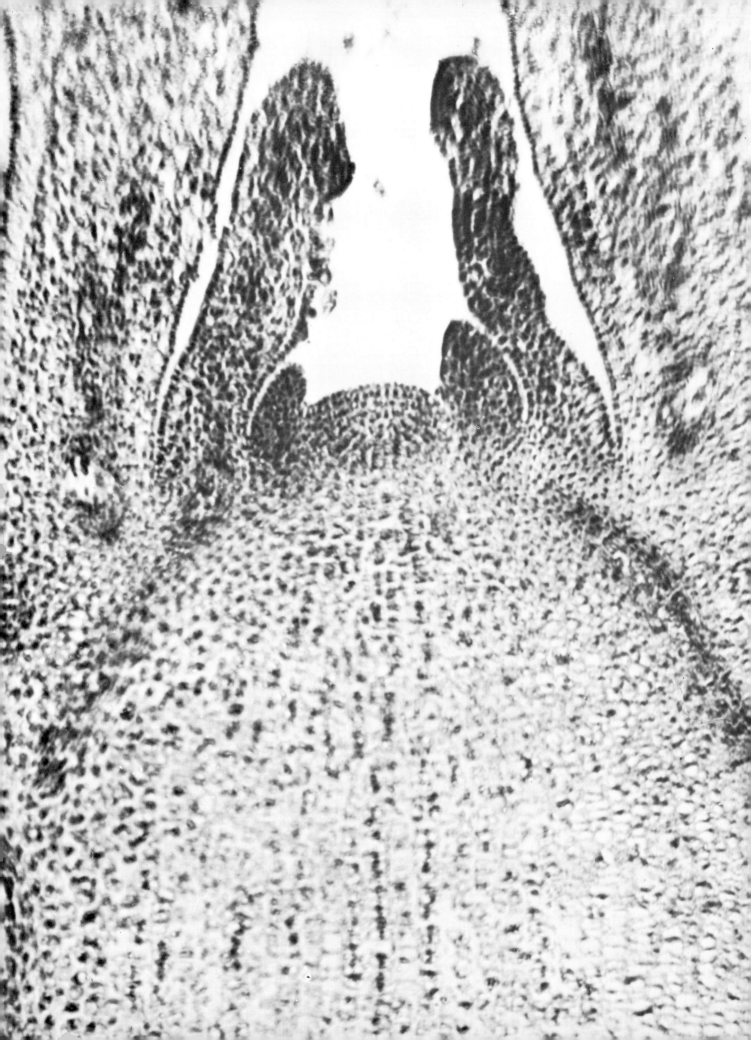

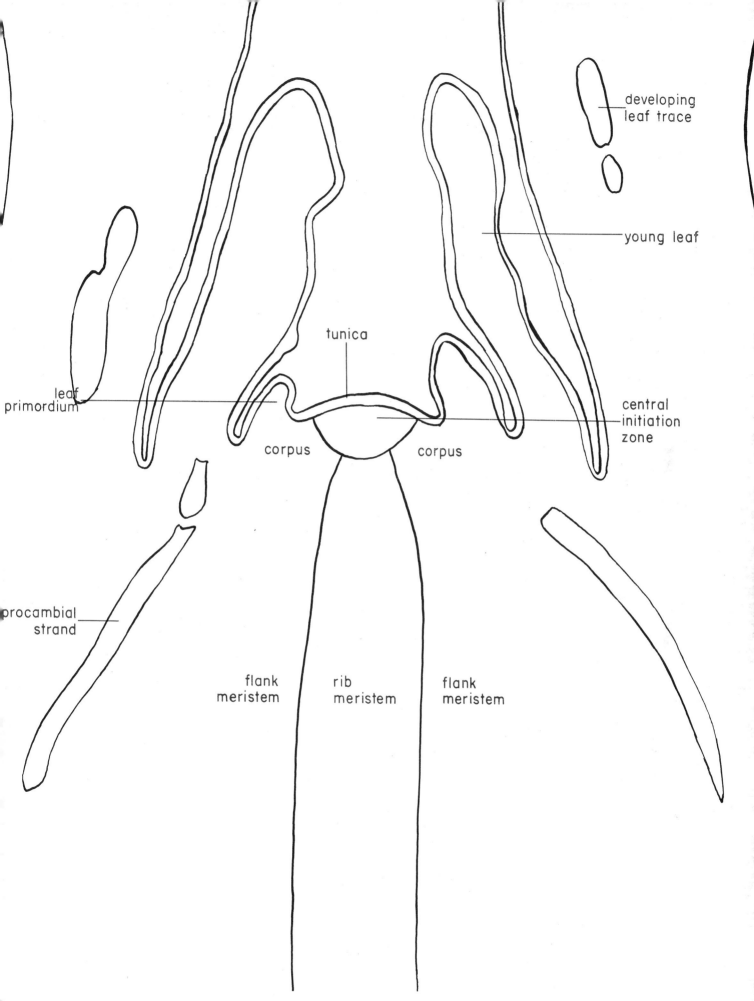

developing
leaf trace

young leaf

tunica

central
initiation
zone

leaf
primordium

corpus

corpus

procambial
strand

flank
meristem

rib
meristem

flank
meristem

Fig. 15.   Low power diagram of a longitudinal section through a stem apex of *Vicia faba*

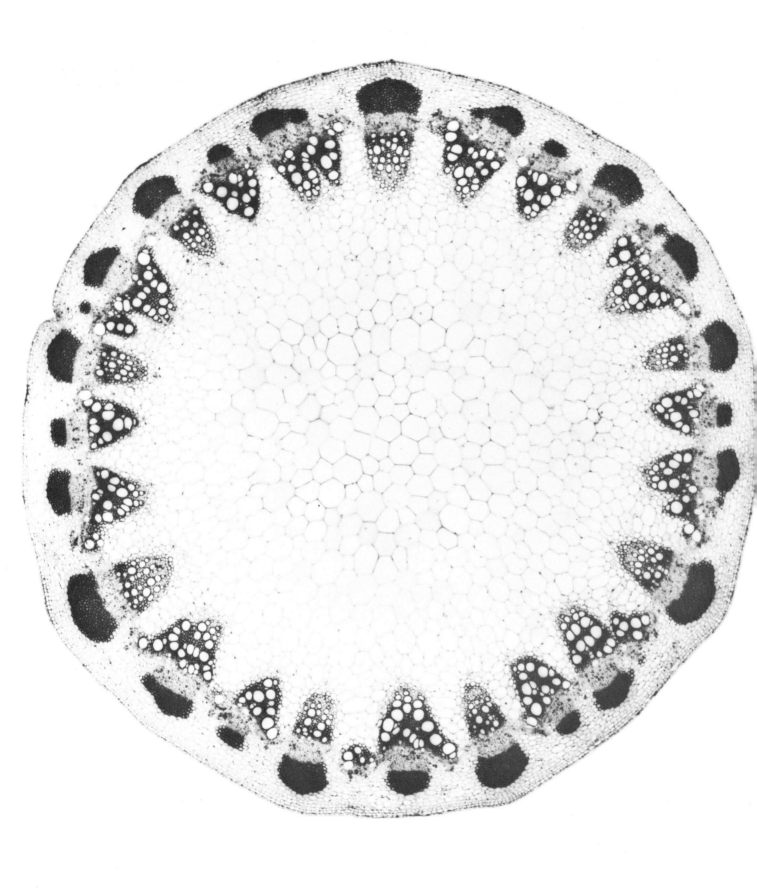

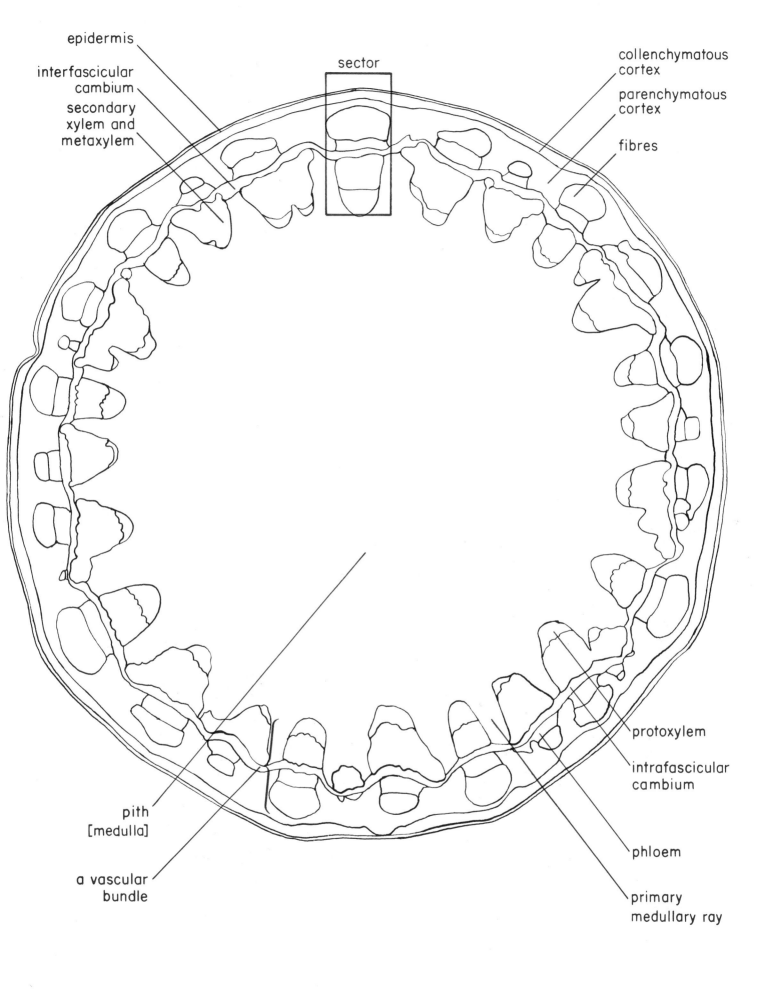

epidermis

interfascicular cambium

secondary xylem and metaxylem

sector

collenchymatous cortex

parenchymatous cortex

fibres

protoxylem

intrafascicular cambium

phloem

primary medullary ray

pith [medulla]

a vascular bundle

Fig. 16.   Low power diagram of a transverse section through a stem of *Helianthus annuus*

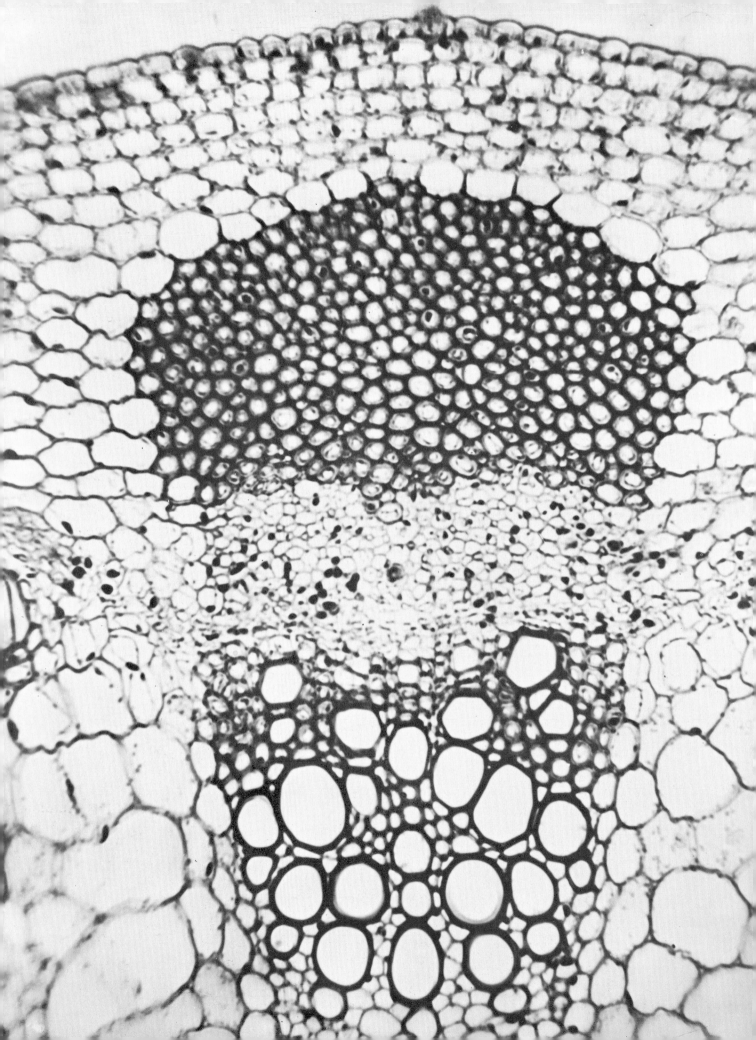

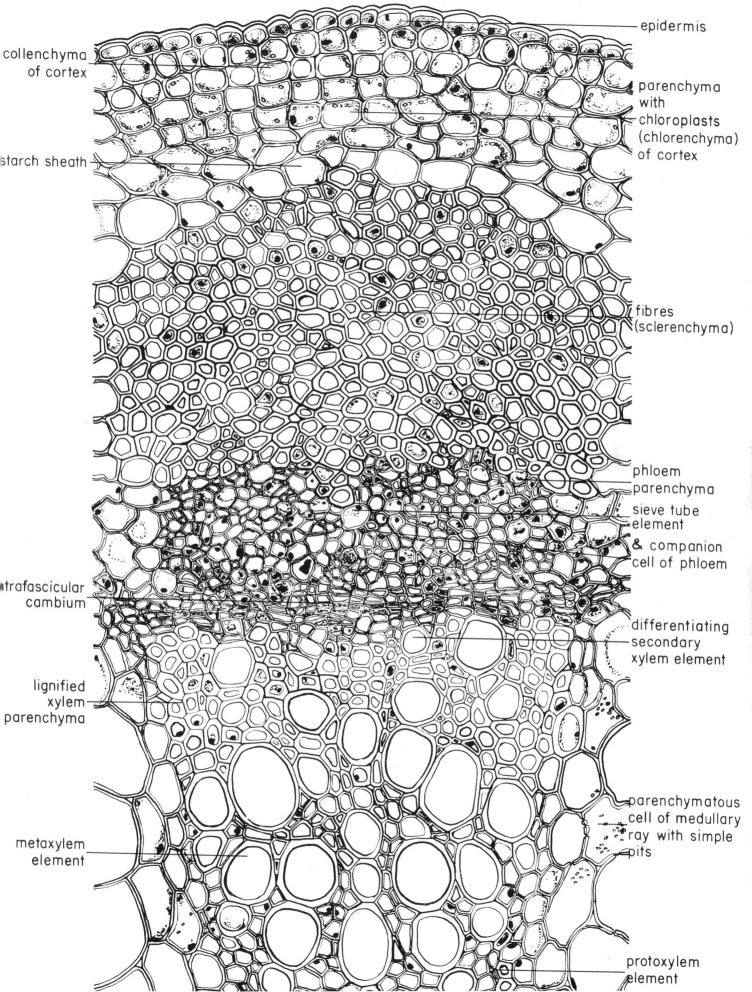

epidermis

collenchyma
of cortex

parenchyma
with
chloroplasts
(chlorenchyma)
of cortex

starch sheath

fibres
(sclerenchyma)

phloem
parenchyma

sieve tube
element

& companion
cell of phloem

intrafascicular
cambium

differentiating
secondary
xylem element

lignified
xylem
parenchyma

parenchymatous
cell of medullary
ray with simple
pits

metaxylem
element

protoxylem
element

Fig. 17.   High power drawing of a sector from a transverse section through a stem of *Helianthus annuus*   39

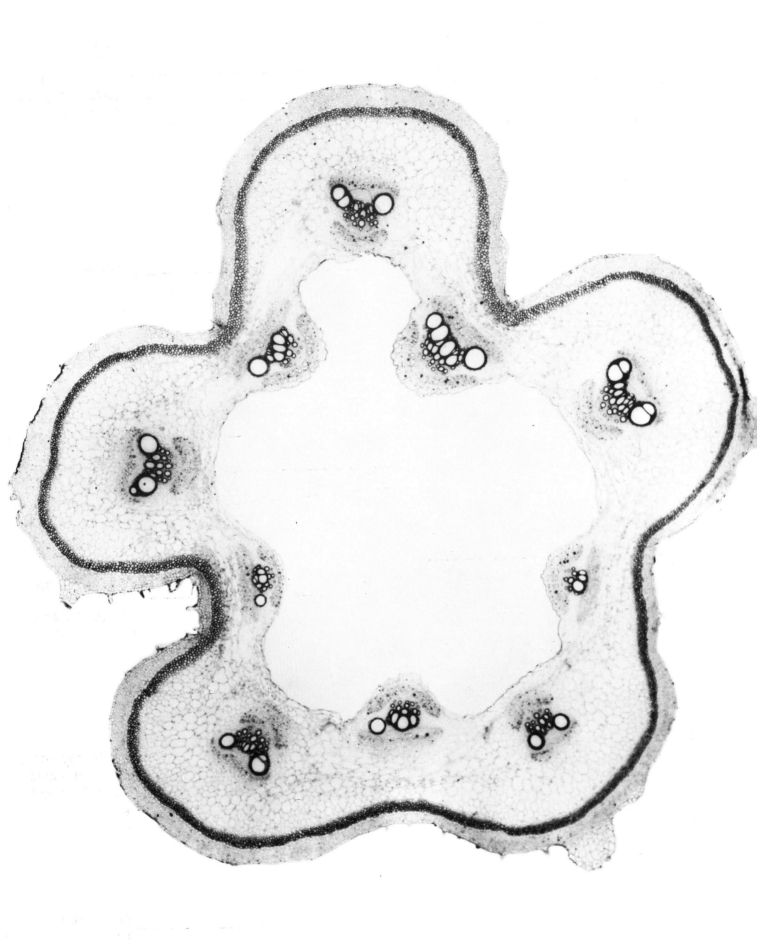

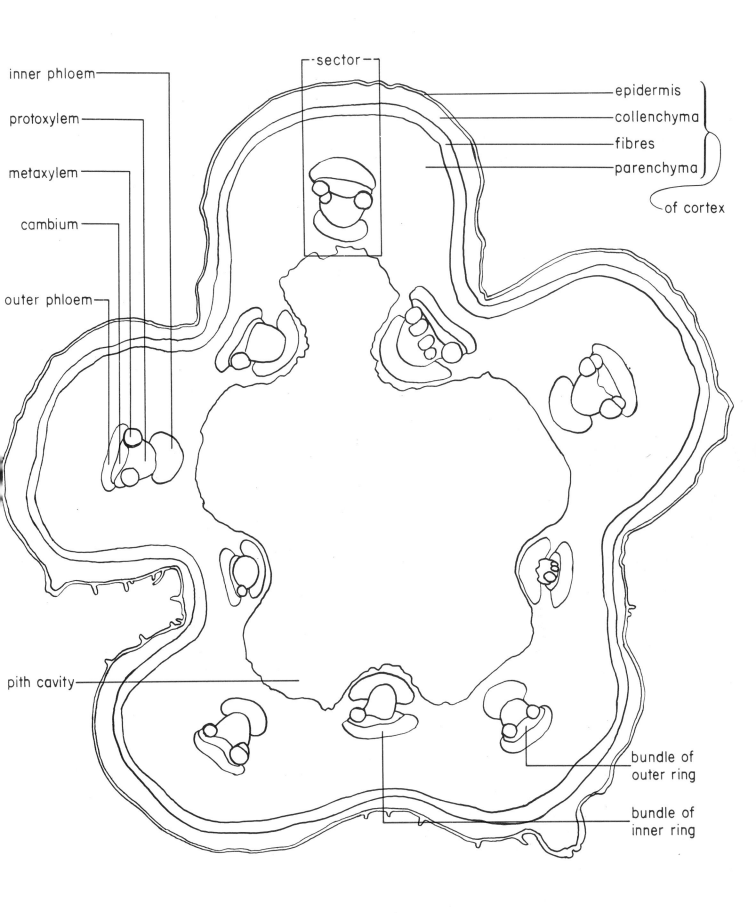

inner phloem

protoxylem

metaxylem

cambium

outer phloem

pith cavity

sector

epidermis

collenchyma

fibres

parenchyma

of cortex

bundle of outer ring

bundle of inner ring

Fig. 18.   Low power diagram of a transverse section through a stem of *Cucurbita pepo*

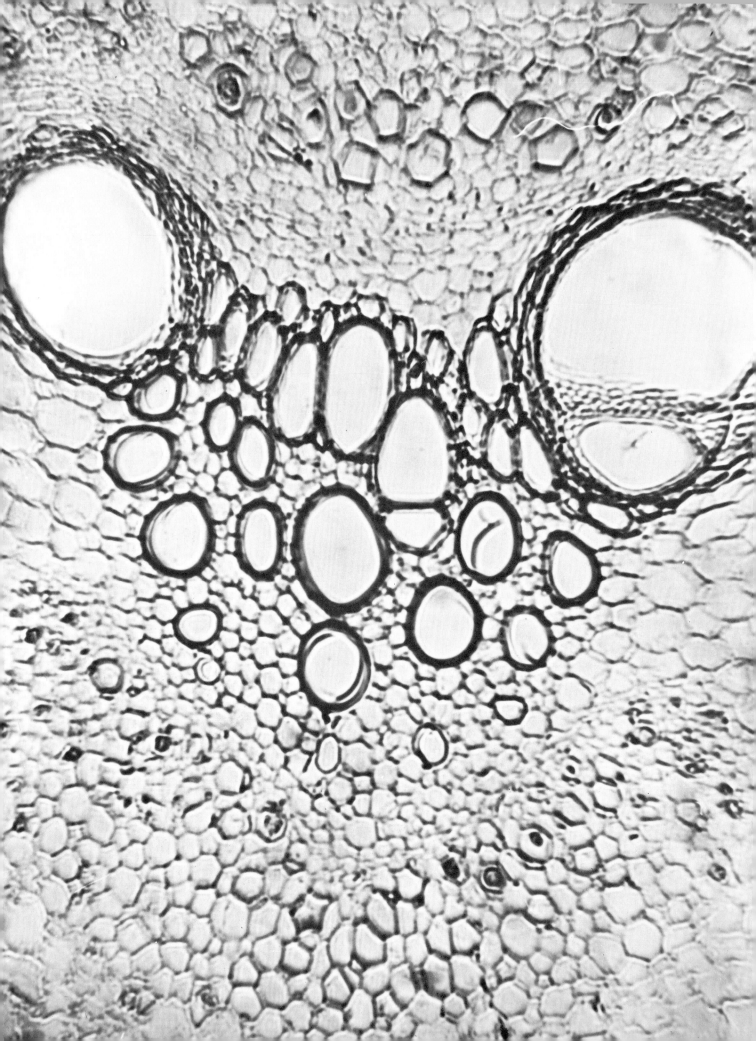

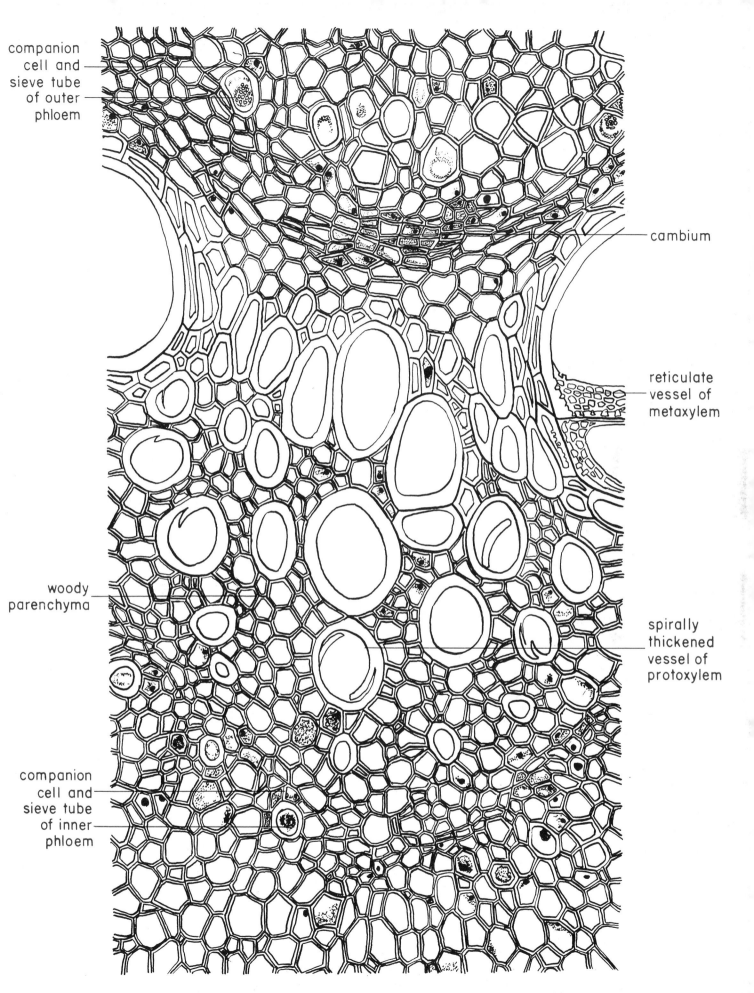

companion cell and sieve tube of outer phloem

cambium

reticulate vessel of metaxylem

woody parenchyma

spirally thickened vessel of protoxylem

companion cell and sieve tube of inner phloem

Fig. 19.   High power drawing of a sector of a transverse section through the stem of *Cucurbita pepo*

43

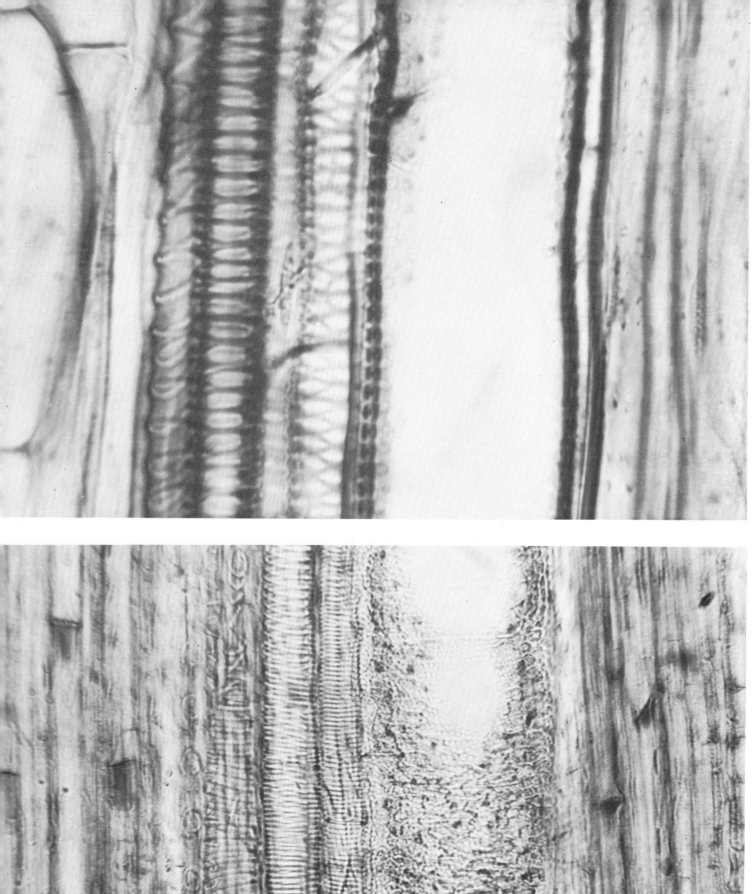

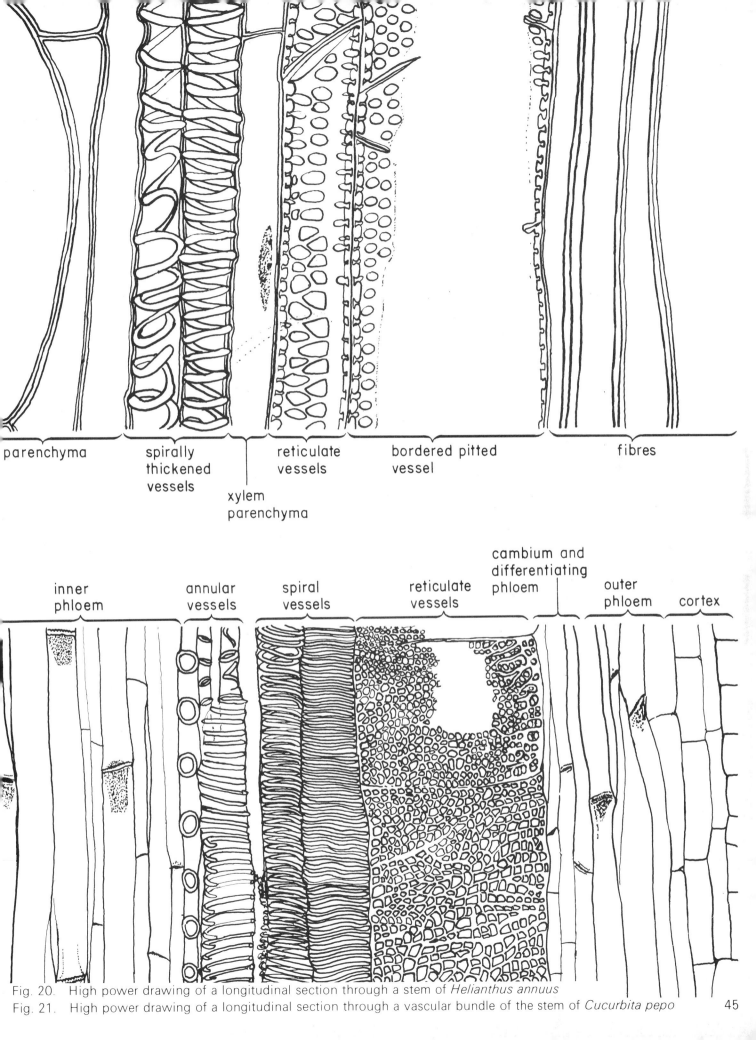

parenchyma   spirally thickened vessels   reticulate vessels   bordered pitted vessel   fibres

xylem parenchyma

inner phloem   annular vessels   spiral vessels   reticulate vessels   cambium and differentiating phloem   outer phloem   cortex

Fig. 20.   High power drawing of a longitudinal section through a stem of *Helianthus annuus*

Fig. 21.   High power drawing of a longitudinal section through a vascular bundle of the stem of *Cucurbita pepo*   45

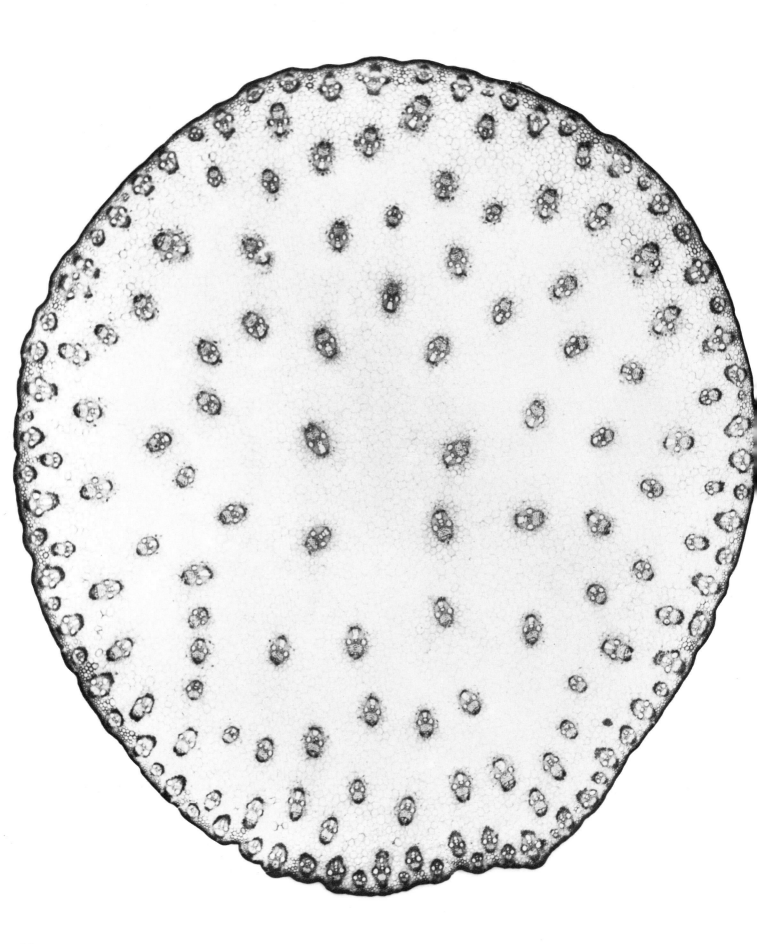

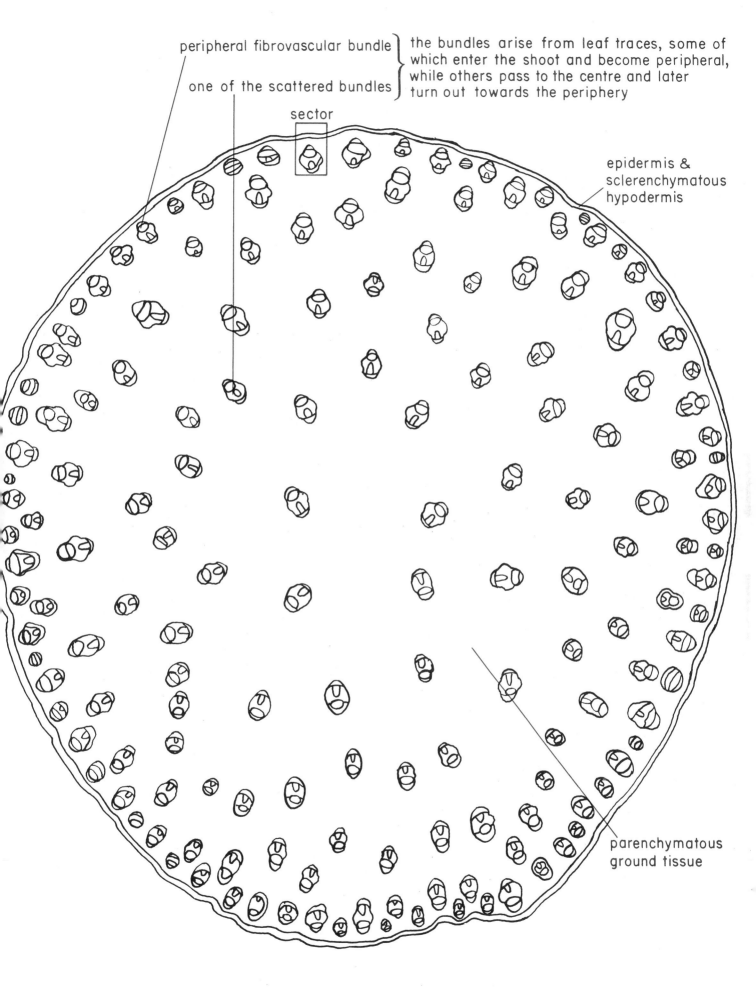

peripheral fibrovascular bundle

one of the scattered bundles

the bundles arise from leaf traces, some of which enter the shoot and become peripheral, while others pass to the centre and later turn out towards the periphery

sector

epidermis & sclerenchymatous hypodermis

parenchymatous ground tissue

Fig. 22.   Low power diagram of a transverse section through a stem of *Zea mais*

47

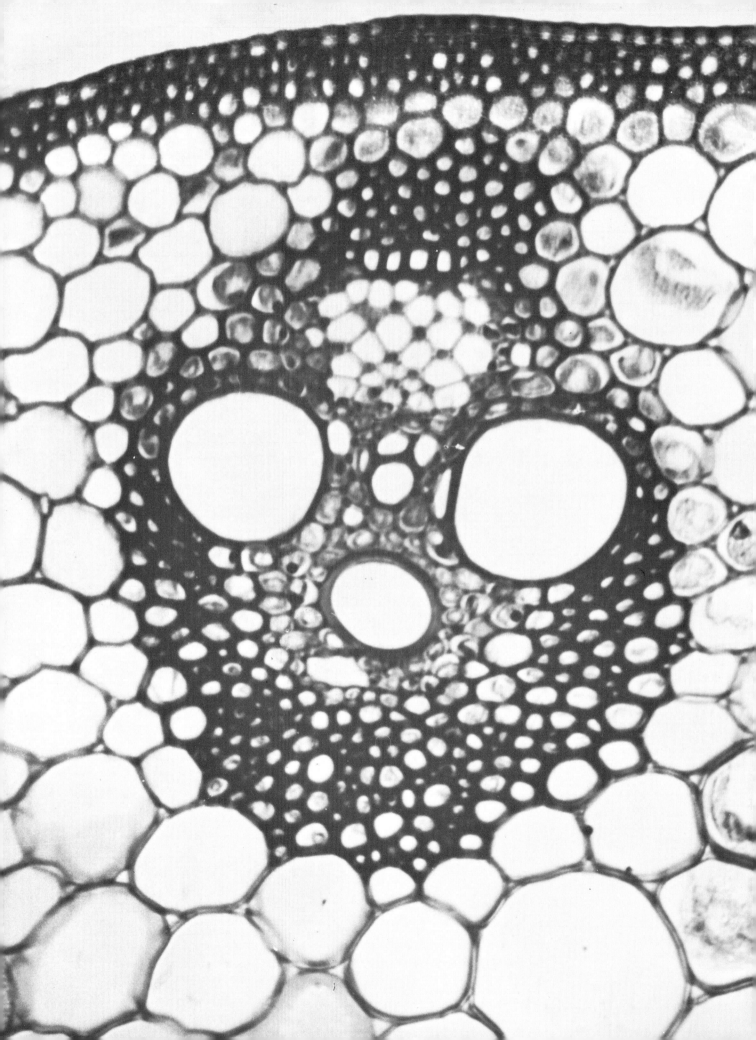

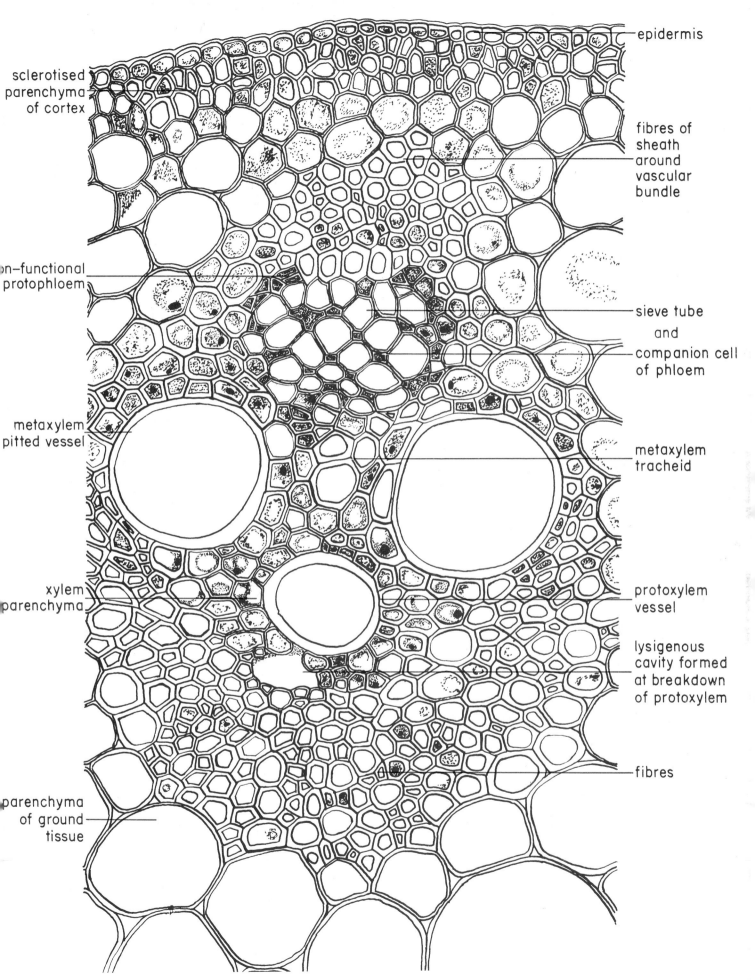

epidermis

sclerotised parenchyma of cortex

fibres of sheath around vascular bundle

on-functional protophloem

sieve tube and

companion cell of phloem

metaxylem pitted vessel

metaxylem tracheid

xylem parenchyma

protoxylem vessel

lysigenous cavity formed at breakdown of protoxylem

fibres

parenchyma of ground tissue

Fig. 23.   High power drawing of a sector from a transverse section through a stem of *Zea mais* to show a fibro-vascular bundle

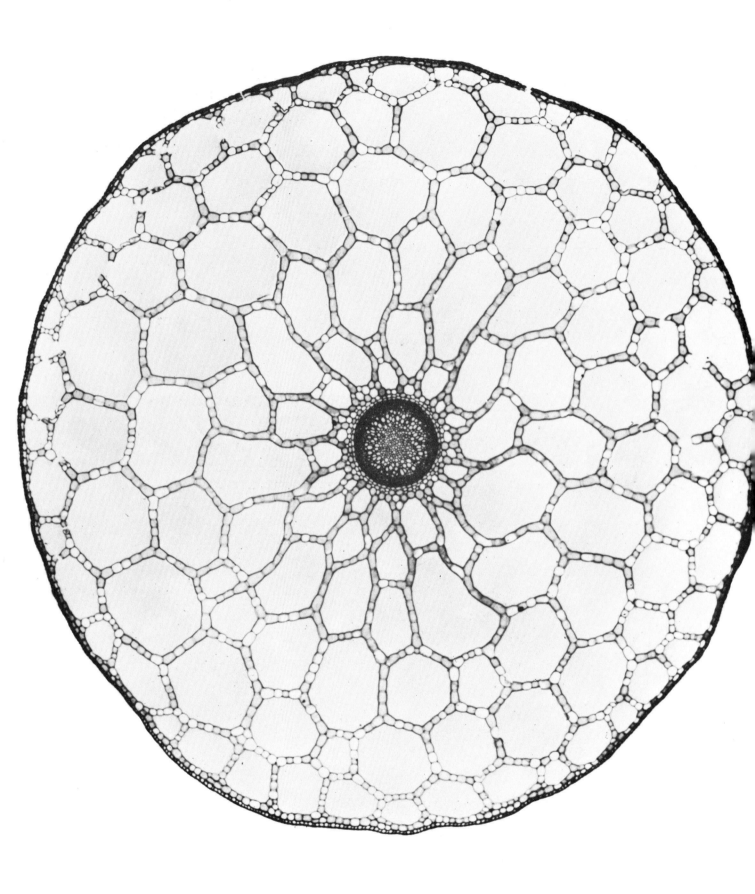

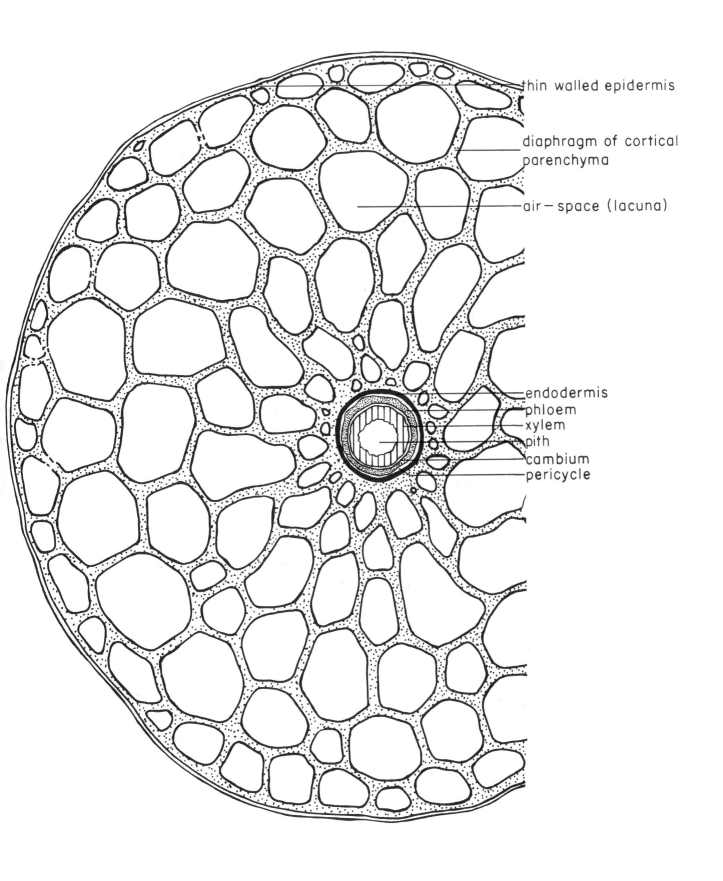

thin walled epidermis

diaphragm of cortical parenchyma

air — space (lacuna)

endodermis
phloem
xylem
pith
cambium
pericycle

Fig. 24.   Low power diagram of a transverse section through a stem of *Hippuris vulgaris*

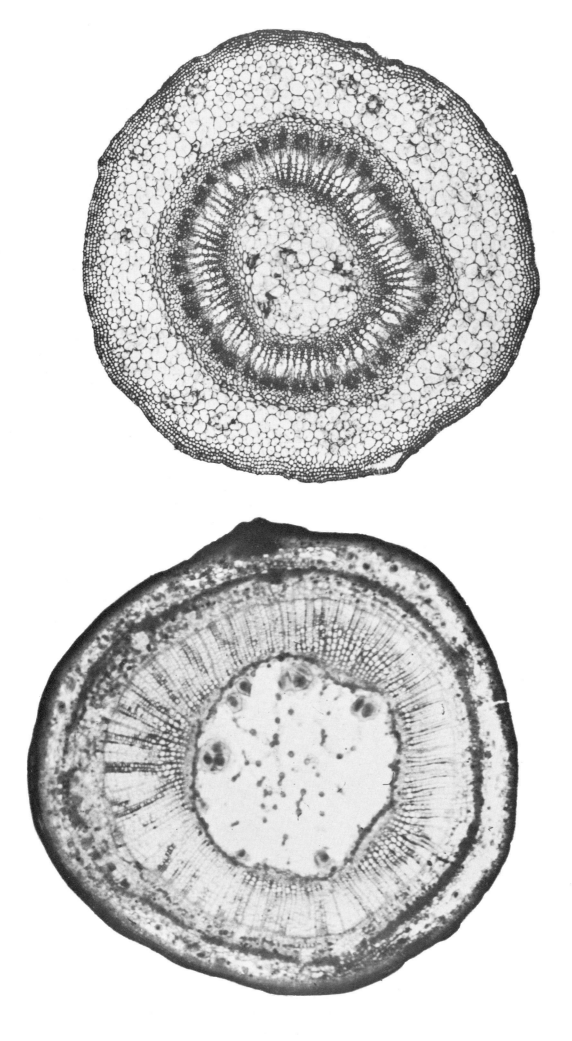

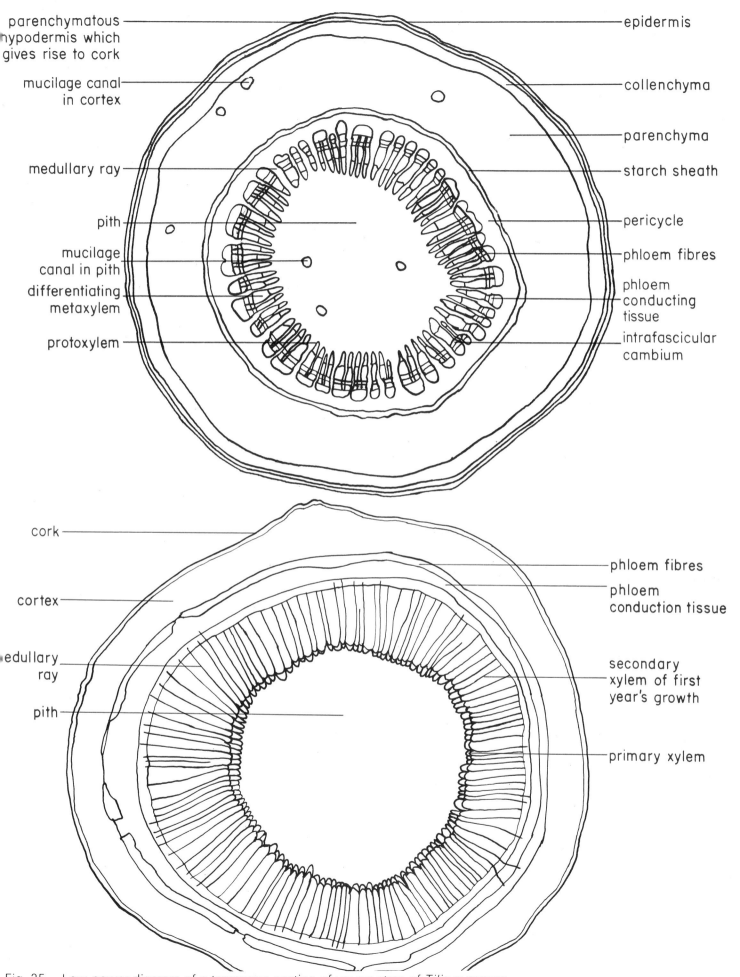

parenthymatous hypodermis which gives rise to cork

epidermis

mucilage canal in cortex

collenchyma

parenchyma

medullary ray

starch sheath

pith

pericycle

mucilage canal in pith

phloem fibres

differentiating metaxylem

phloem conducting tissue

protoxylem

intrafascicular cambium

cork

phloem fibres

cortex

phloem conduction tissue

edullary ray

secondary xylem of first year's growth

pith

primary xylem

Fig. 25. Low power diagram of a transverse section of young stem of *Tilia europaea*

Fig. 26. Low power diagram of a transverse section through a one-year old stem of *Tilia europaea*

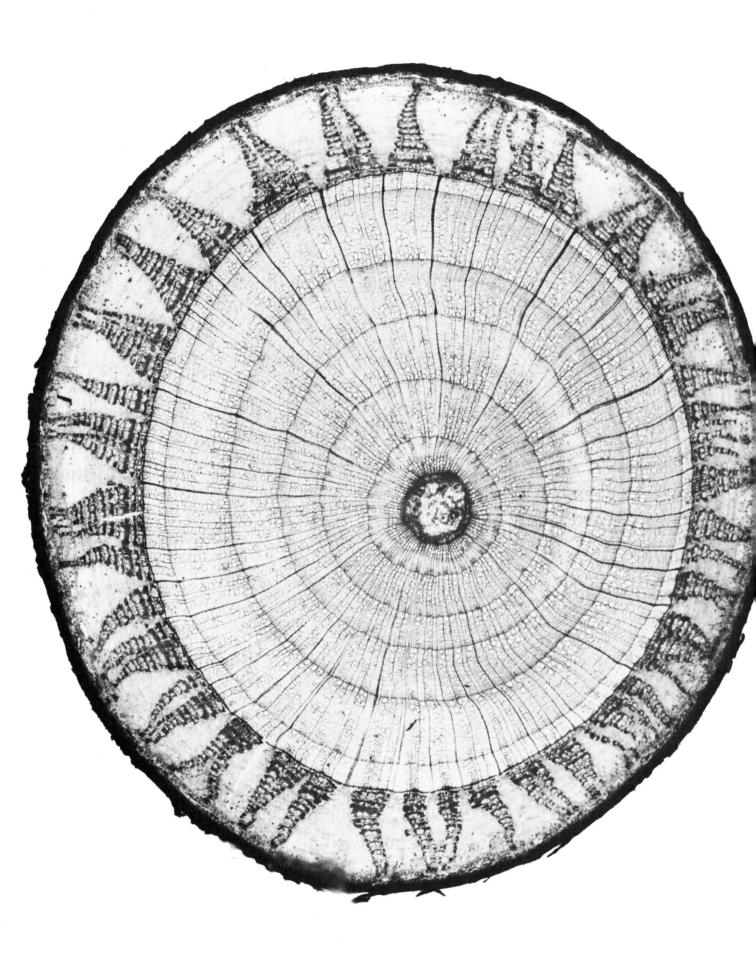

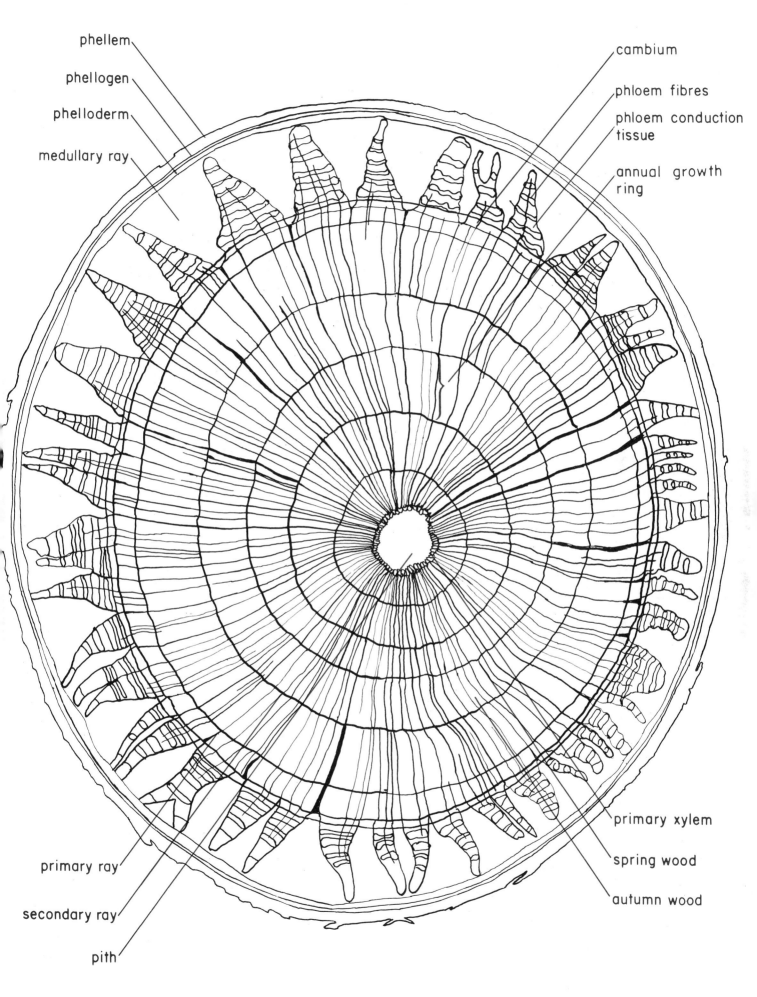

phellem

phellogen

phelloderm

medullary ray

cambium

phloem fibres

phloem conduction tissue

annual growth ring

primary xylem

spring wood

autumn wood

primary ray

secondary ray

pith

Fig. 27. Low power diagram of a transverse section through a six-year old stem of *Tilia europaea*   55

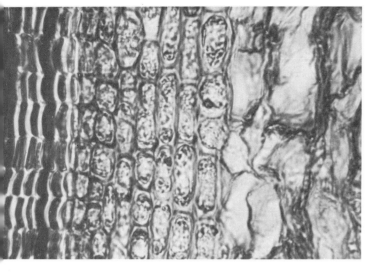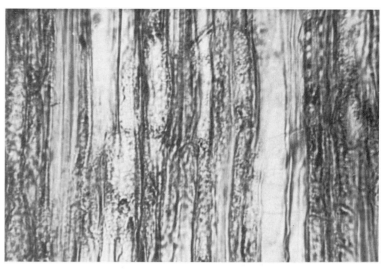

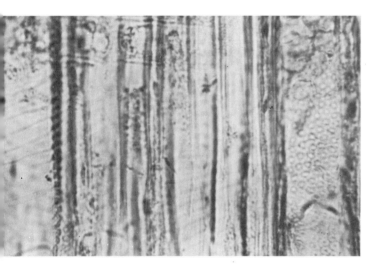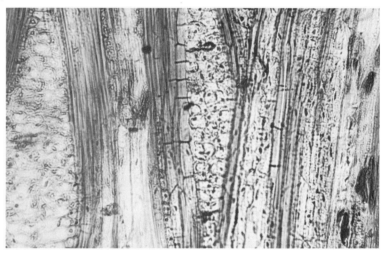

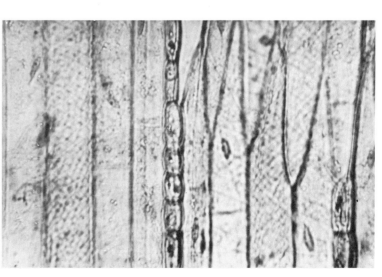

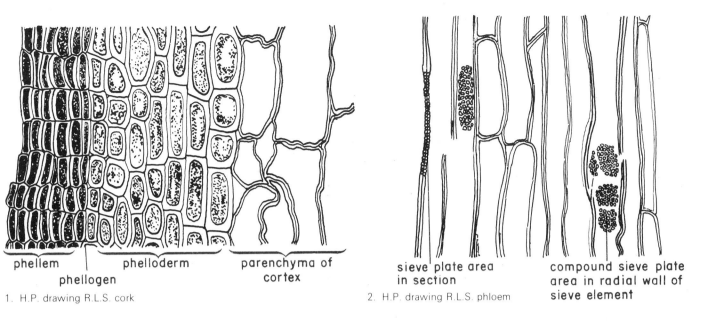

1. H.P. drawing R.L.S. cork

phellem | phellogen | phelloderm | parenchyma of cortex

2. H.P. drawing R.L.S. phloem

sieve plate area in section | compound sieve plate area in radial wall of sieve element

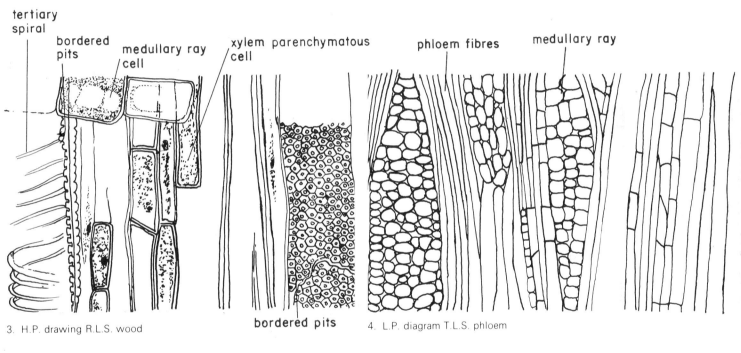

tertiary spiral | bordered pits | medullary ray cell | xylem parenchymatous cell

bordered pits

3. H.P. drawing R.L.S. wood

phloem fibres | medullary ray

4. L.P. diagram T.L.S. phloem

5. H.P. drawing T.L.S. phloem

fibre | medullary ray | compound sieve plate

6. H.P. drawing T.L.S. wood

pitted vessel | uniseriate ray | bordered pitted tracheids with tertiary spirals

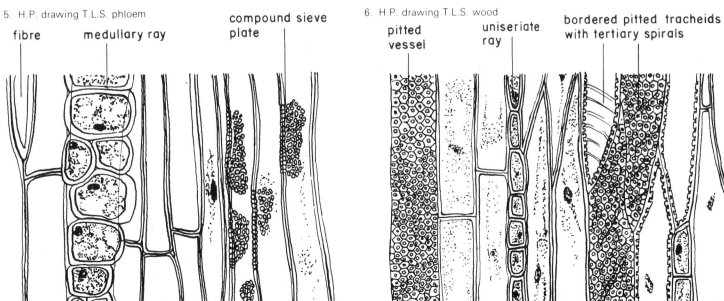

Fig. 28. Longitudinal sections of *Tilia europaea* stem

57

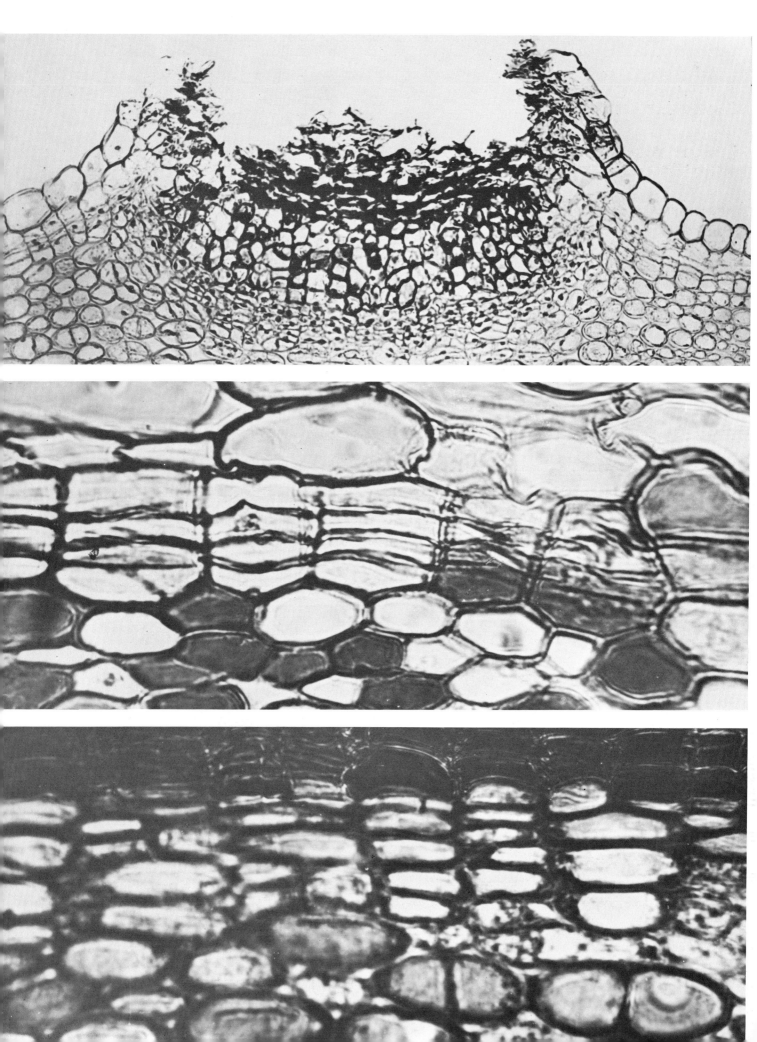

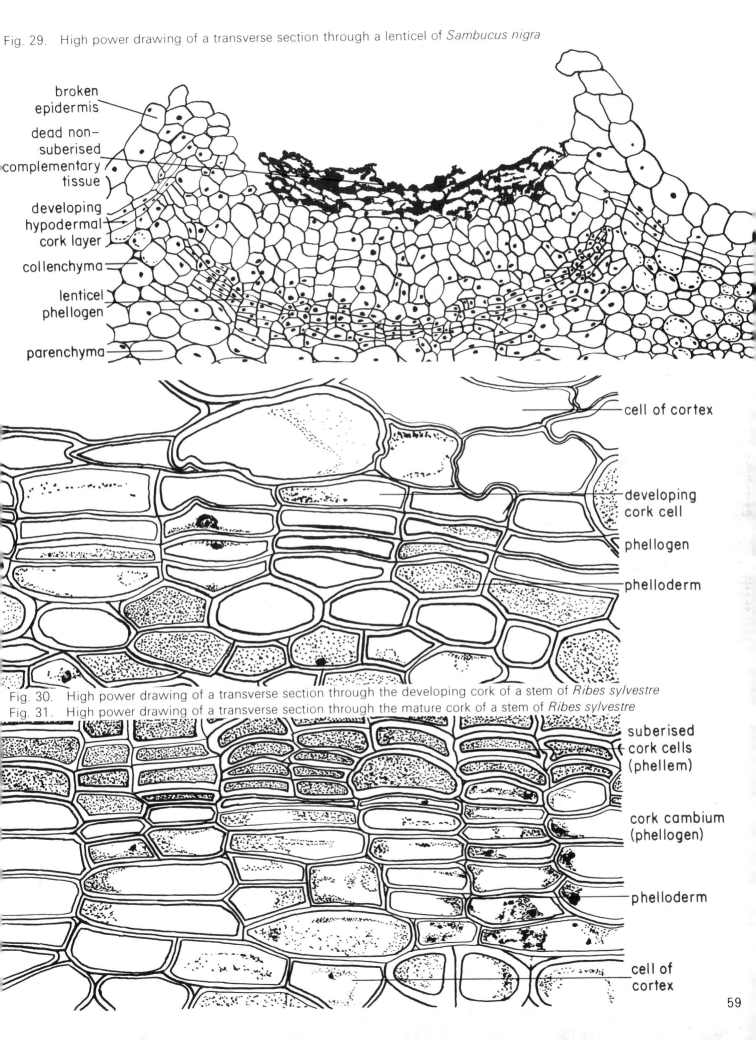

Fig. 29. High power drawing of a transverse section through a lenticel of *Sambucus nigra*

broken epidermis

dead non-suberised complementary tissue

developing hypodermal cork layer

collenchyma

lenticel phellogen

parenchyma

cell of cortex

developing cork cell

phellogen

phelloderm

Fig. 30. High power drawing of a transverse section through the developing cork of a stem of *Ribes sylvestre*
Fig. 31. High power drawing of a transverse section through the mature cork of a stem of *Ribes sylvestre*

suberised cork cells (phellem)

cork cambium (phellogen)

phelloderm

cell of cortex

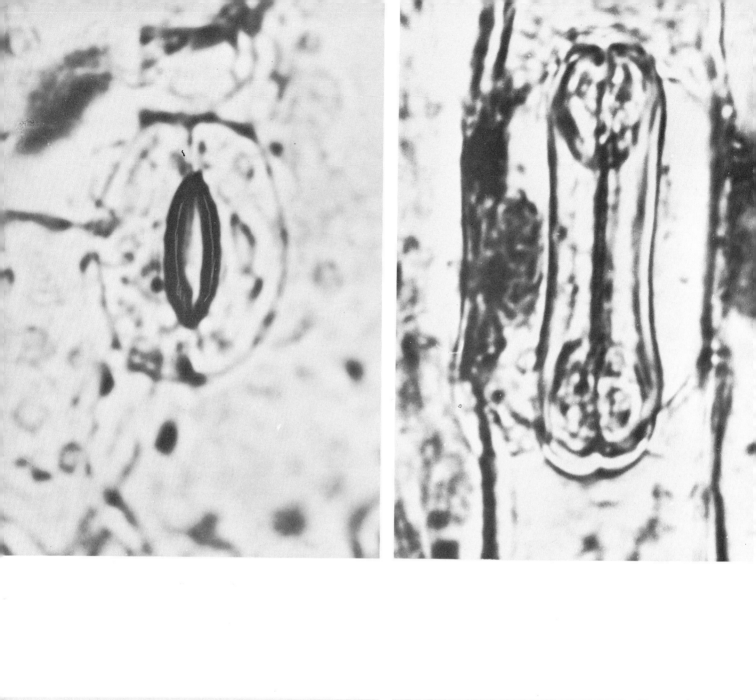
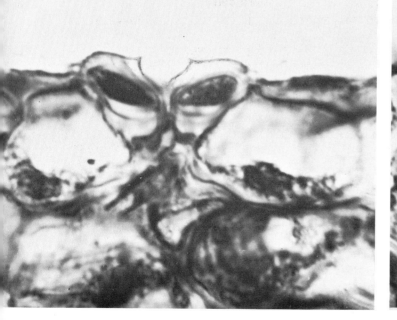
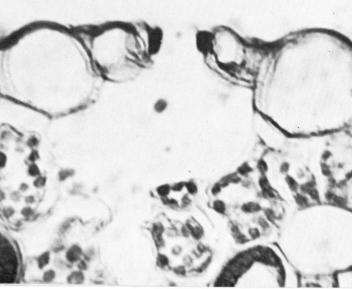

Fig. 32. High power drawing of a stoma in surface view from a leaf of *Vicia faba*
Fig. 34. High power drawing of a stoma in surface view from a leaf of
Dactylis glomerata

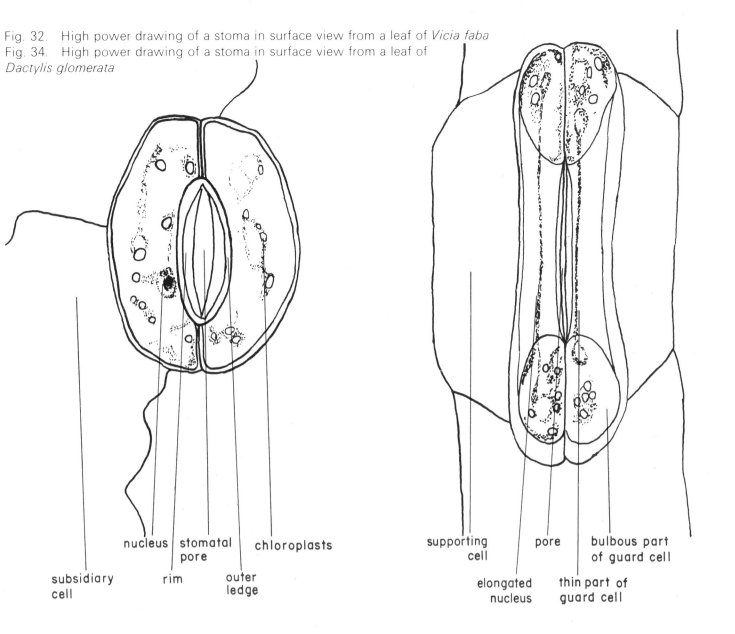

nucleus   stomatal   chloroplasts
           pore

subsidiary   rim   outer
cell   ledge

supporting   pore   bulbous part
cell   of guard cell

elongated   thin part of
nucleus   guard cell

Fig. 33. High power drawing of a transverse section through a stoma of a leaf of *Prunus laurocerasus*
Fig. 35. High power drawing of a transverse section through a stoma of a leaf of *Zea mais*

thick cuticle   outer ledge   guard cell

subsidiary cell   pore   heavy cutinisation   respiratory chamber   subsidiary cell
                          of guard cell wall

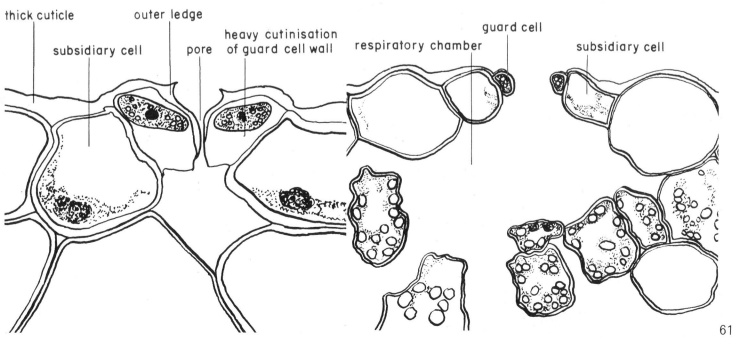

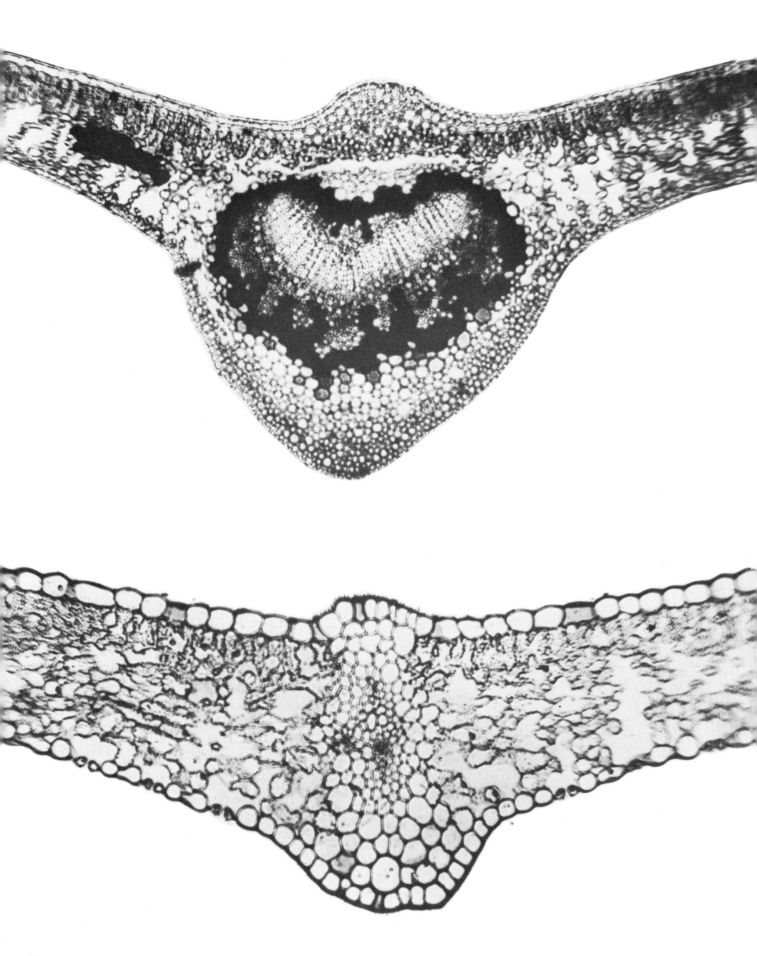

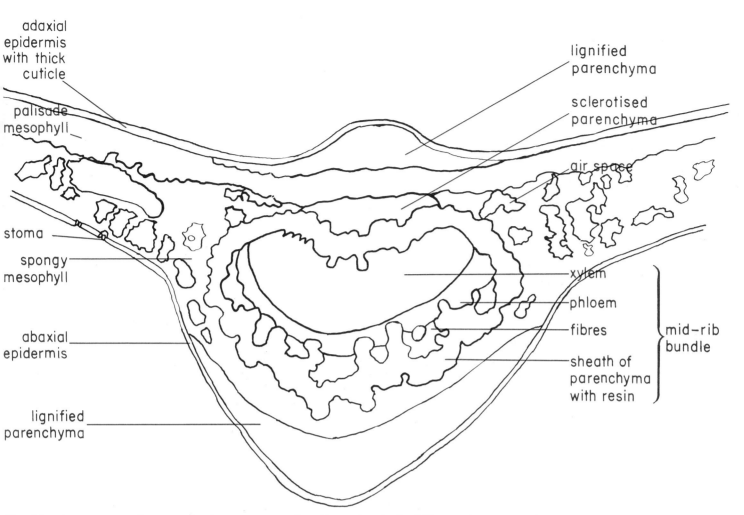

adaxial epidermis with thick cuticle

palisade mesophyll

stoma

spongy mesophyll

abaxial epidermis

lignified parenchyma

lignified parenchyma

sclerotised parenchyma

air space

xylem

phloem

fibres

sheath of parenchyma with resin

} mid-rib bundle

Fig. 36.   Low power diagram of a transverse section through a leaf of *Prunus laurocerasus*

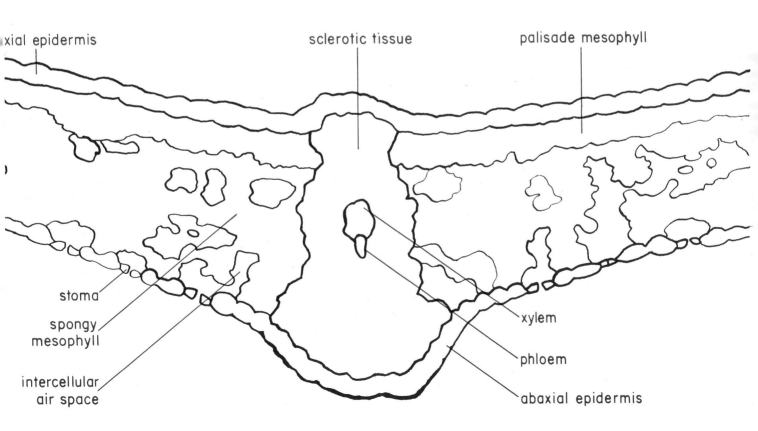

xial epidermis

sclerotic tissue

palisade mesophyll

stoma

spongy mesophyll

intercellular air space

xylem

phloem

abaxial epidermis

Fig. 37.   Low power diagram of a transverse section through a leaf of *Lilium*

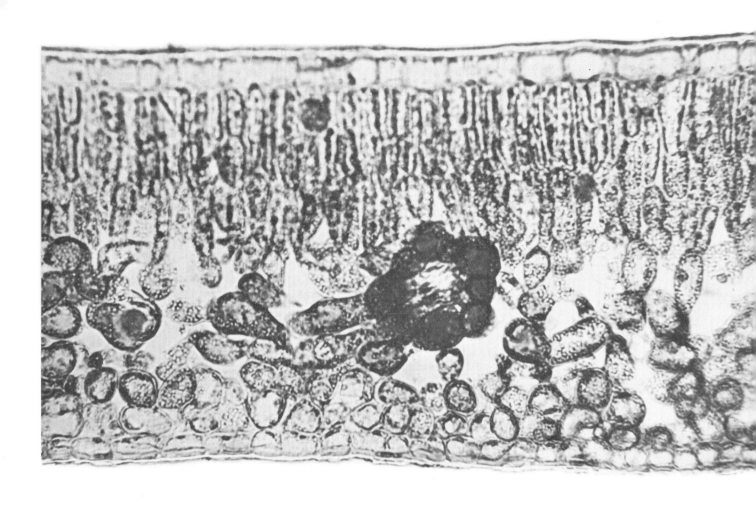

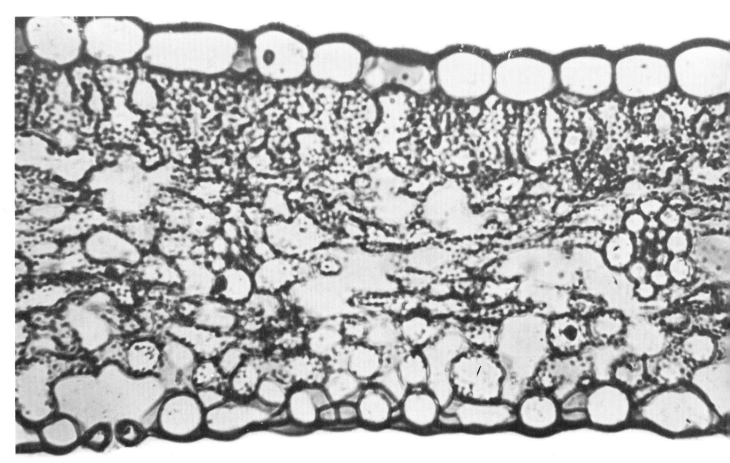

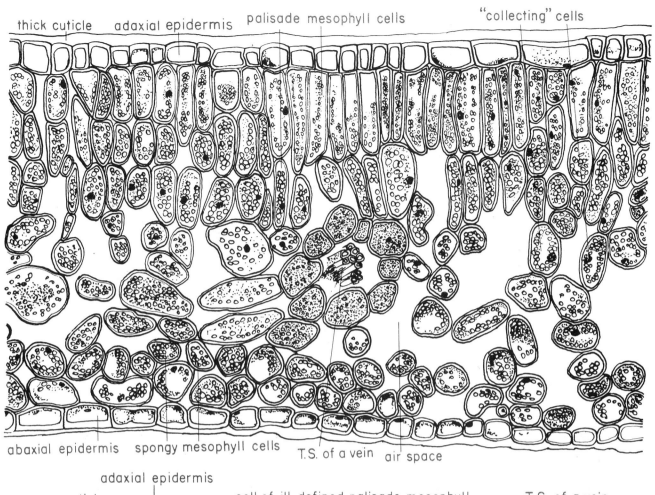

thick cuticle   adaxial epidermis   palisade mesophyll cells   "collecting" cells

abaxial epidermis   spongy mesophyll cells   T.S. of a vein   air space

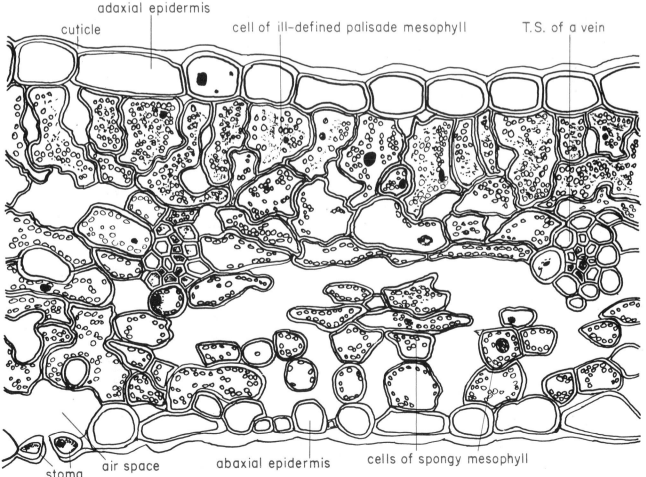

adaxial epidermis

cuticle   cell of ill-defined palisade mesophyll   T.S. of a vein

stoma   air space   abaxial epidermis   cells of spongy mesophyll

Fig. 38.   High power drawing of a transverse section through a leaf of *Prunus laurocerasus*

Fig. 39.   High power drawing of a part of a transverse section of a leaf of *Lilium*

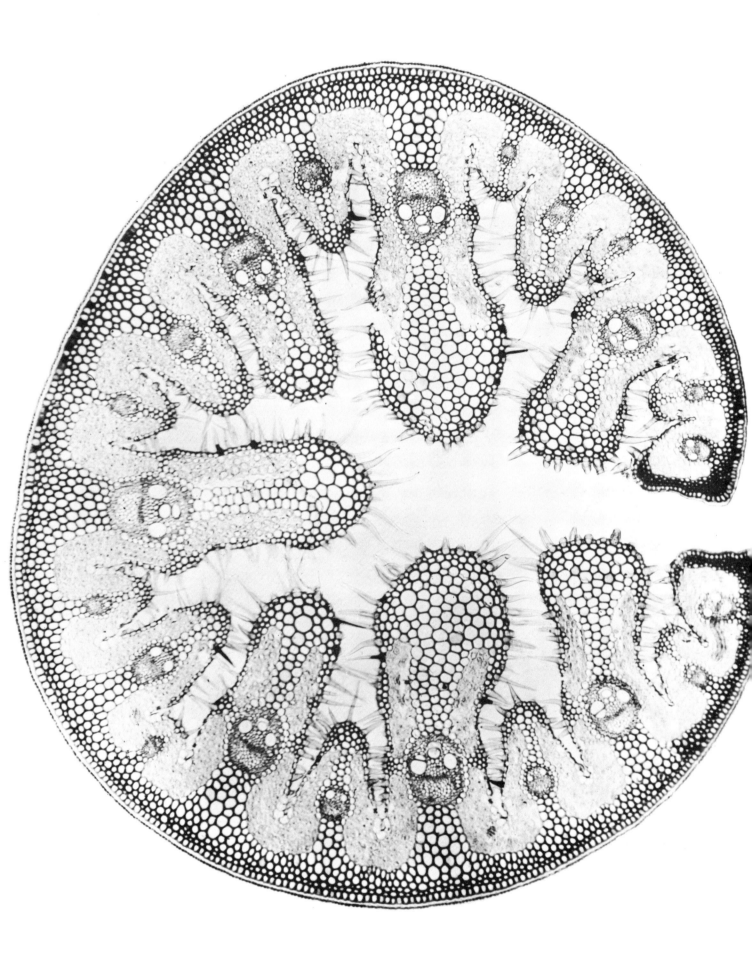

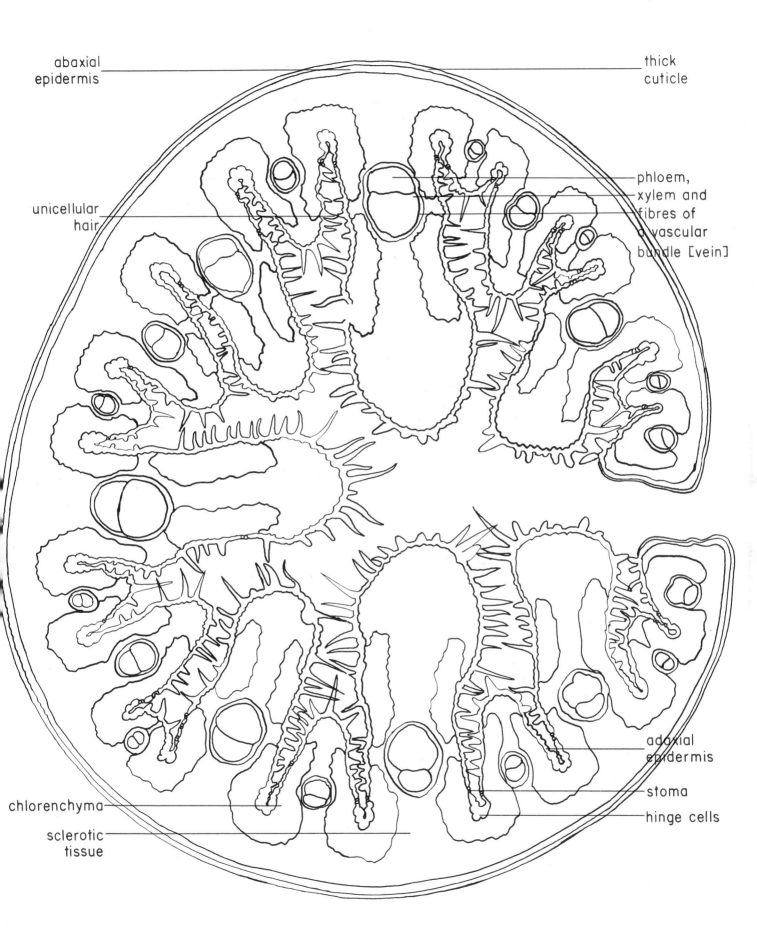

abaxial
epidermis

thick
cuticle

unicellular
hair

phloem,
xylem and
fibres of
a vascular
bundle [vein]

adaxial
epidermis

stoma

hinge cells

chlorenchyma

sclerotic
tissue

Fig. 40.   Low power diagram of a transverse section through a leaf of *Ammophila* [*Psamma*] *arenaria*

67

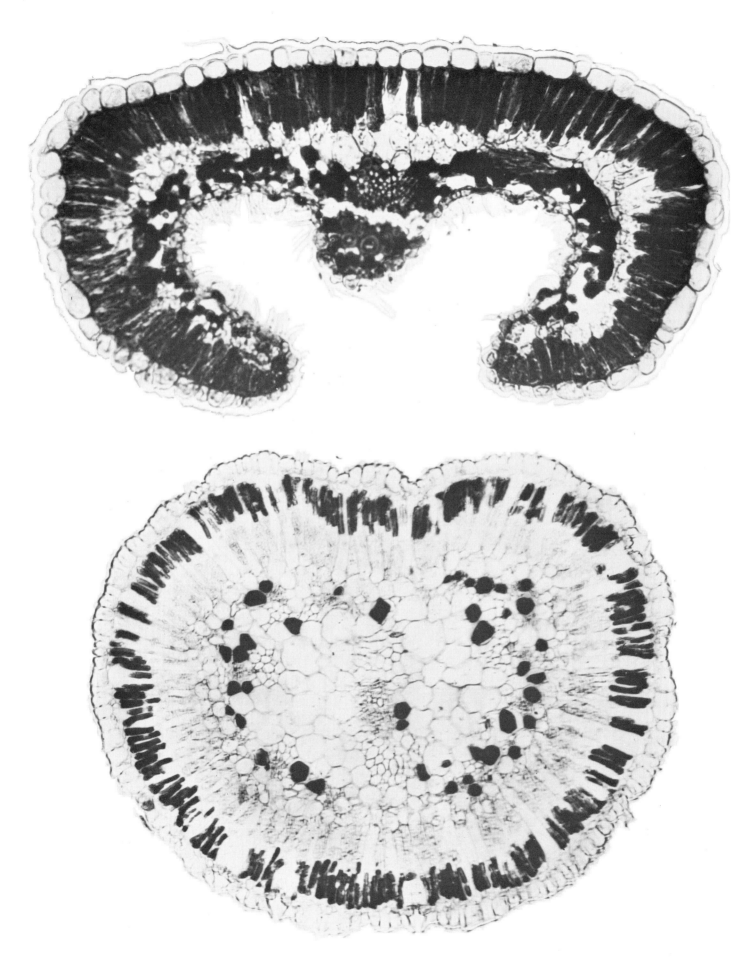

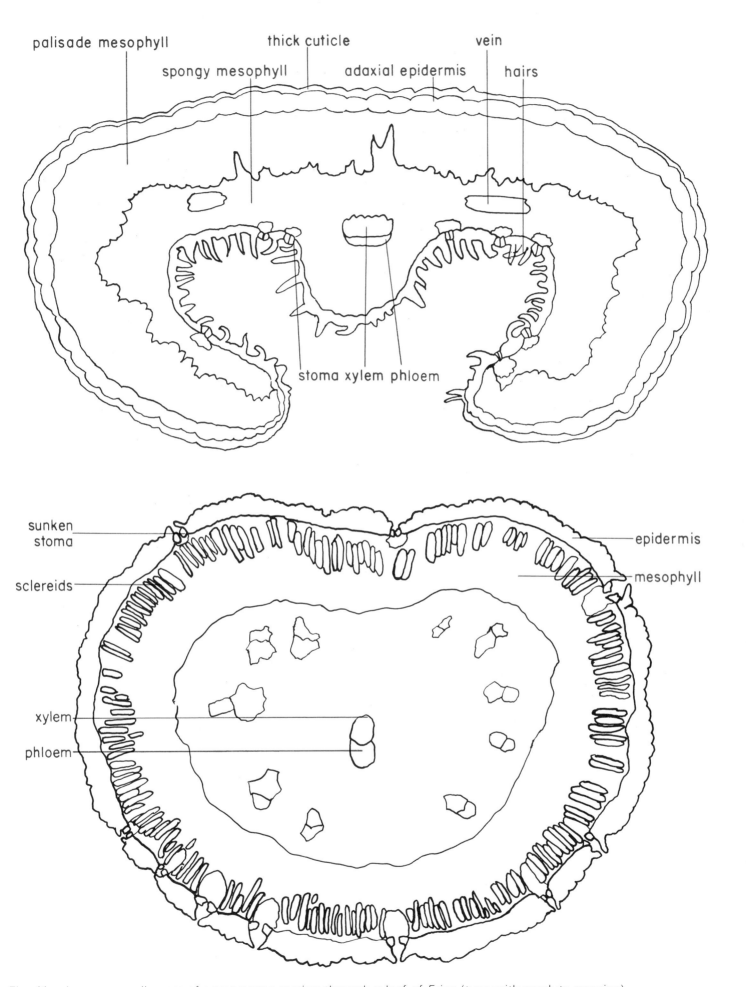

Fig. 41. Low power diagram of a transverse section through a leaf of *Erica* (type with revolute margins)

Fig. 42. Low power diagram of a transverse section through a leaf of *Hakea* (centric arrangement)

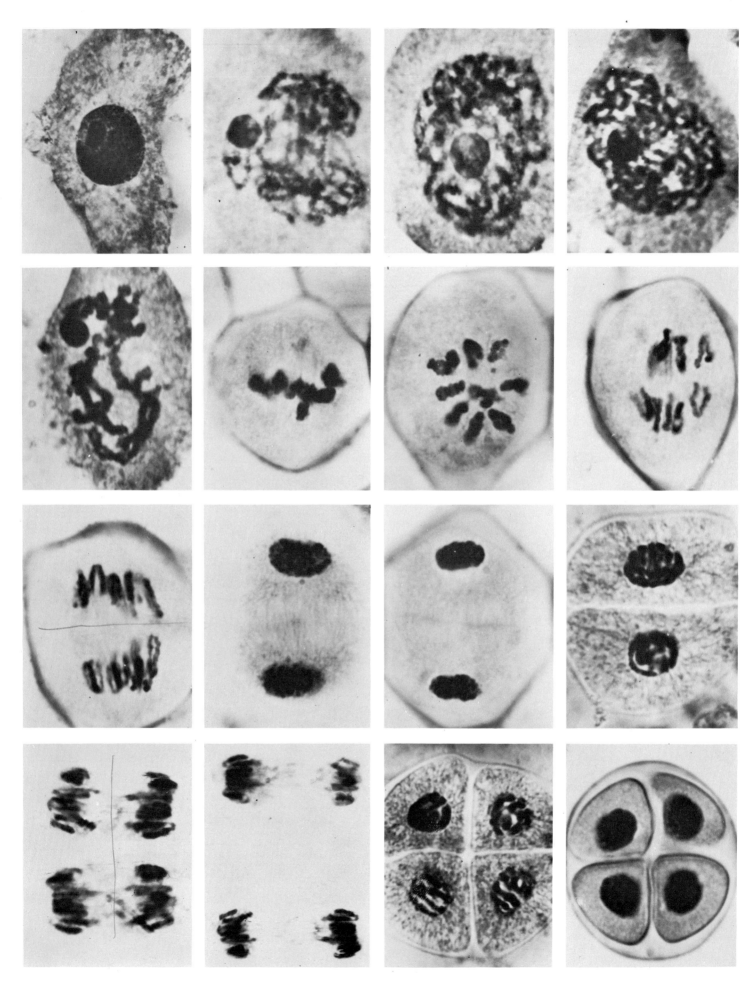

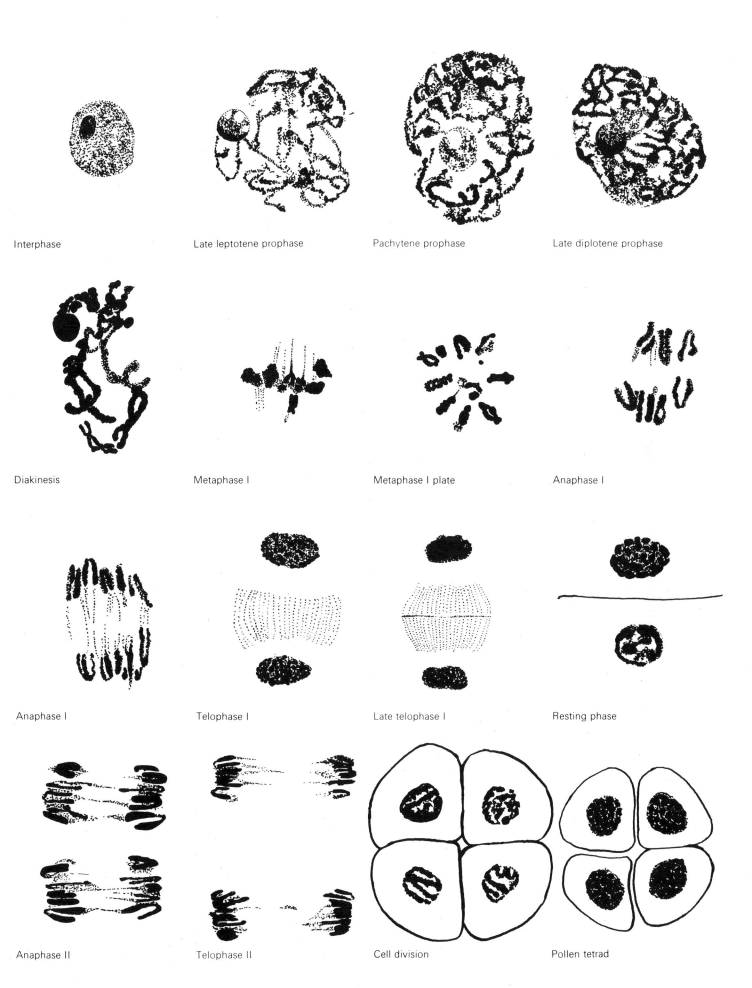

Interphase Late leptotene prophase Pachytene prophase Late diplotene prophase

Diakinesis Metaphase I Metaphase I plate Anaphase I

Anaphase I Telophase I Late telophase I Resting phase

Anaphase II Telophase II Cell division Pollen tetrad

Fig. 43. High power (oil immersion) drawings to show meiosis during microspore development in *Lilium*

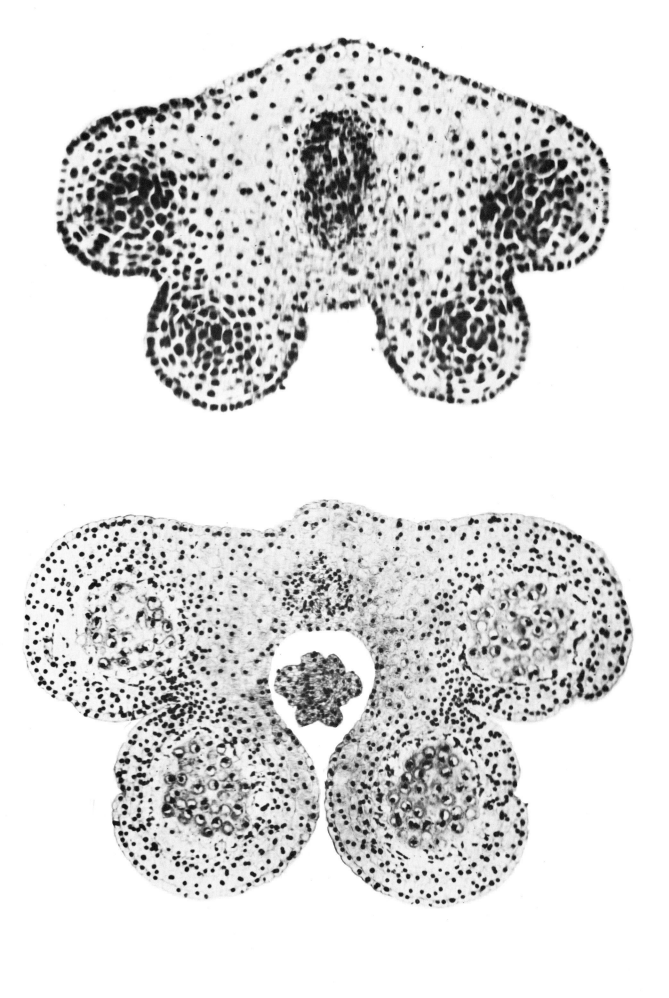

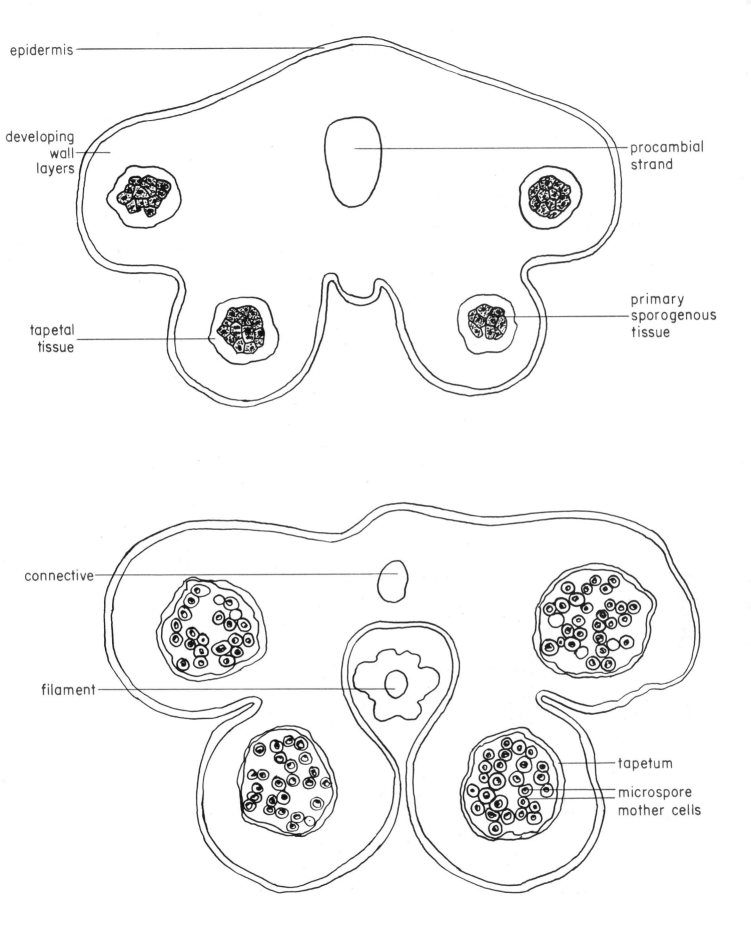

epidermis

developing
wall
layers

procambial
strand

tapetal
tissue

primary
sporogenous
tissue

connective

filament

tapetum

microspore
mother cells

Fig. 44.   Low power diagram of a transverse section through a young anther of *Lilium*
Fig. 45.   Low power diagram of a transverse section through an anther of *Lilium* at a later stage

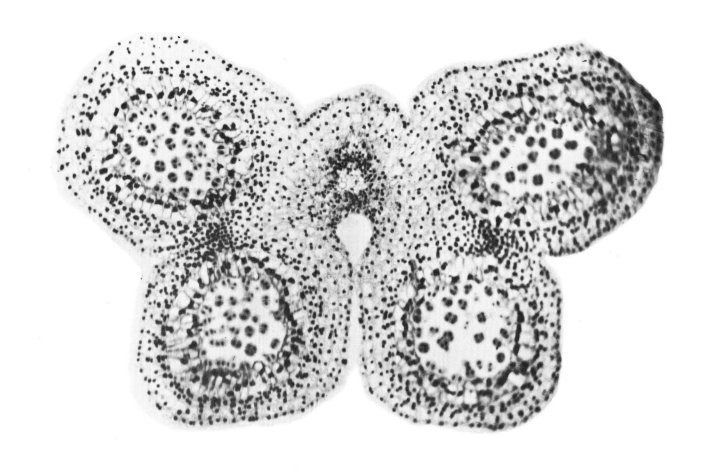

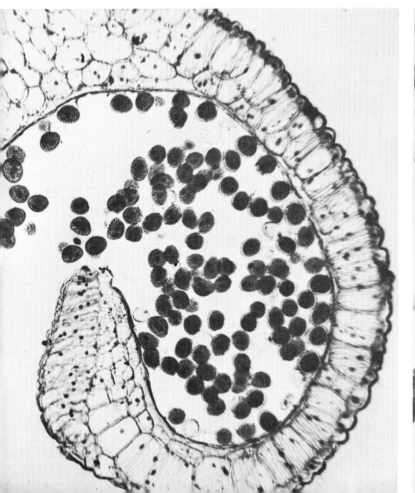

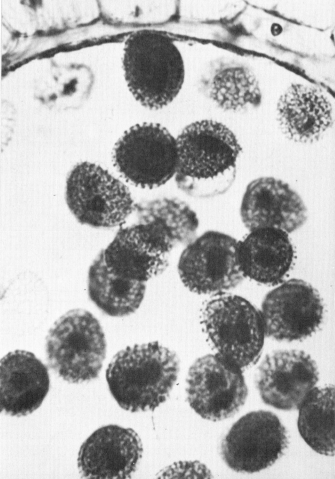

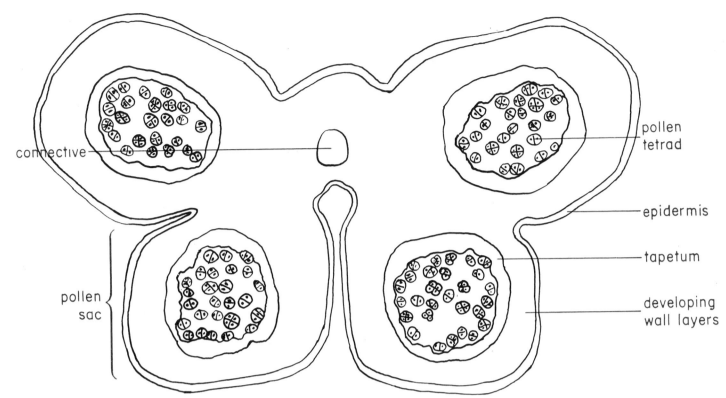

Fig 46. Low power diagram of a transverse section through an unripe anther of *Lilium*

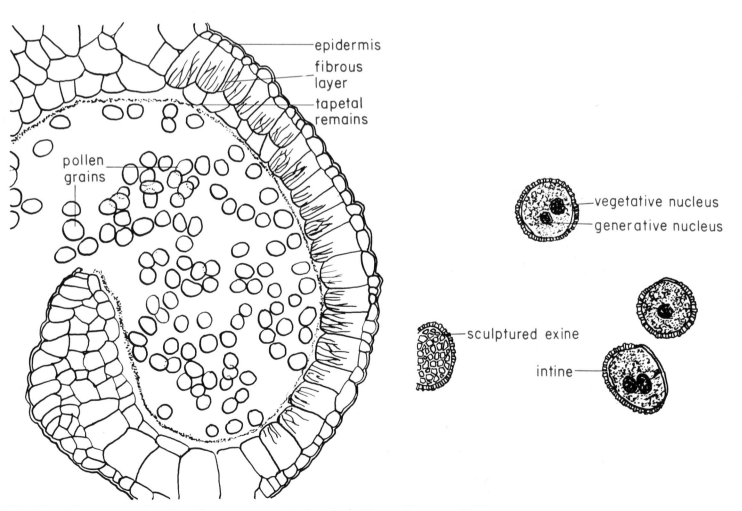

Fig. 47.   High power drawing of a transverse section through a pollen sac of *Lilium*
Fig. 48.   High power (oil immersion) drawing of pollen grains of *Lilium*

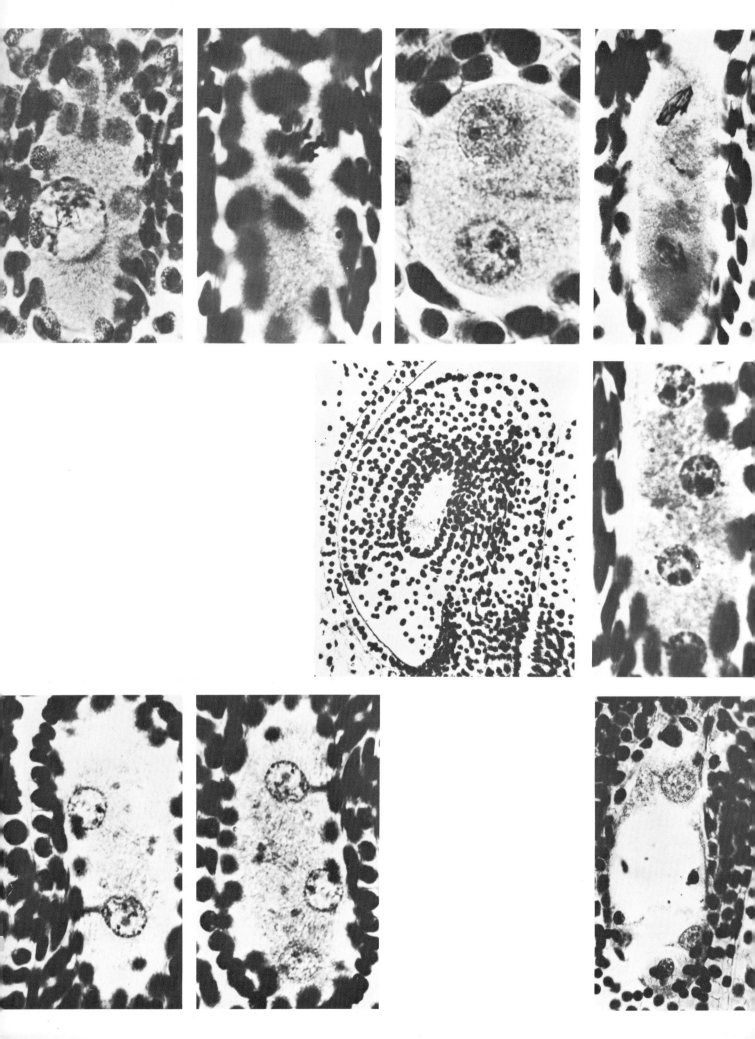

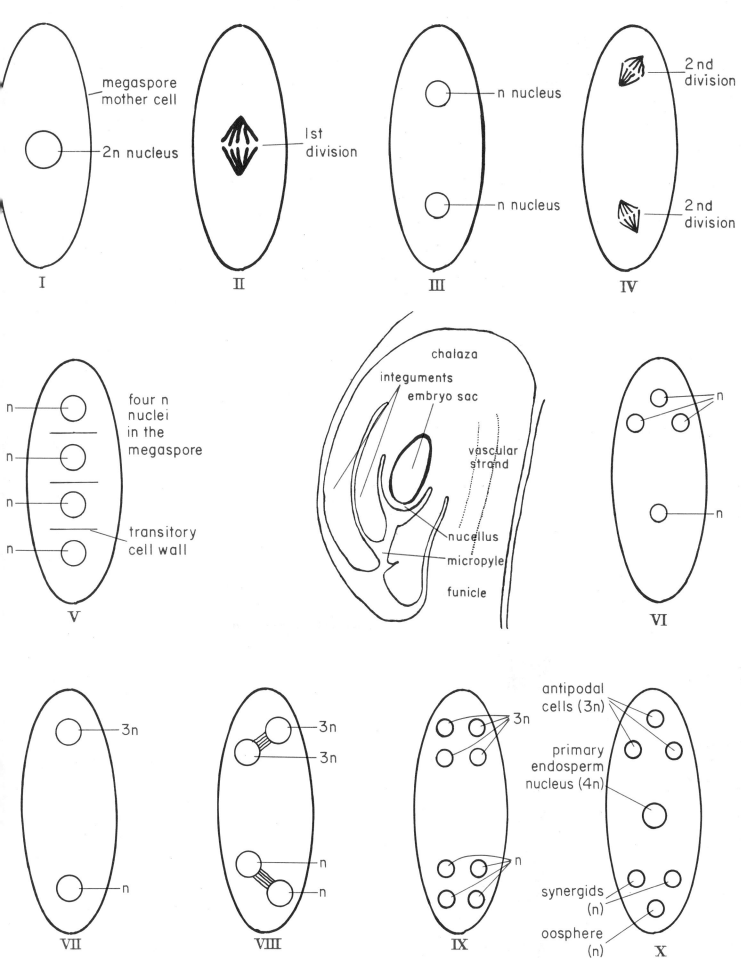

Fig. 49.   One high power drawing and ten diagrams to show the development of the embryo sac of *Lilium*

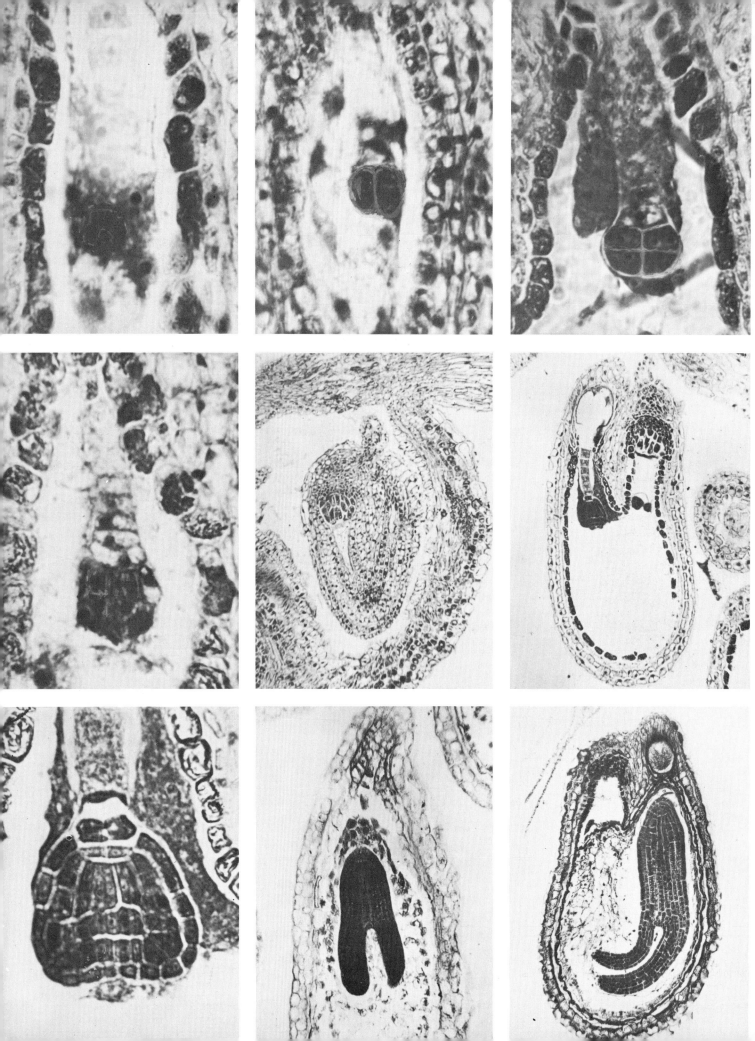

Fig. 50. High power drawings to show the development of the embryo of *Capsella bursa-pastoris*

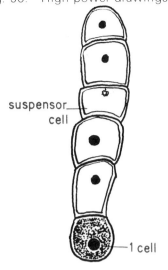

suspensor cell

1 cell

H.P. one celled stage

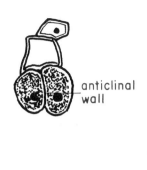

anticlinal wall

H.P. two celled stage

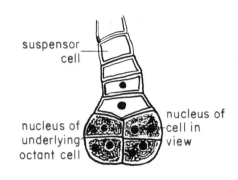

suspensor cell

nucleus of underlying octant cell

nucleus of cell in view

H.P. octet stage

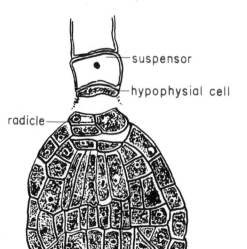

basal cell

cell of suspensor

periclinal wall

H.P. later stage

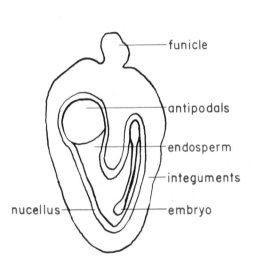

funicle

antipodals

endosperm

integuments

nucellus

embryo

L.P. L.S. of ovule with young embryo

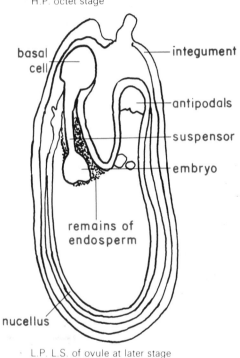

basal cell

integument

antipodals

suspensor

embryo

remains of endosperm

nucellus

L.P. L.S. of ovule at later stage

suspensor

hypophysial cell

radicle

plumule initials

central cylinder initials

H.P. to show differentiation

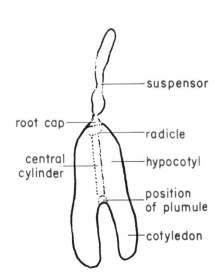

suspensor

root cap

central cylinder

radicle

hypocotyl

position of plumule

cotyledon

L.P. of embryo with developing cotyledons

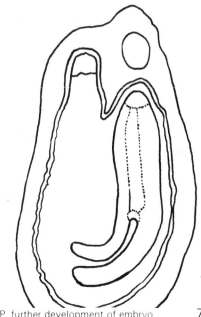

L.P. further development of embryo

Fig. 51. Development of *Capsella* embryo continued

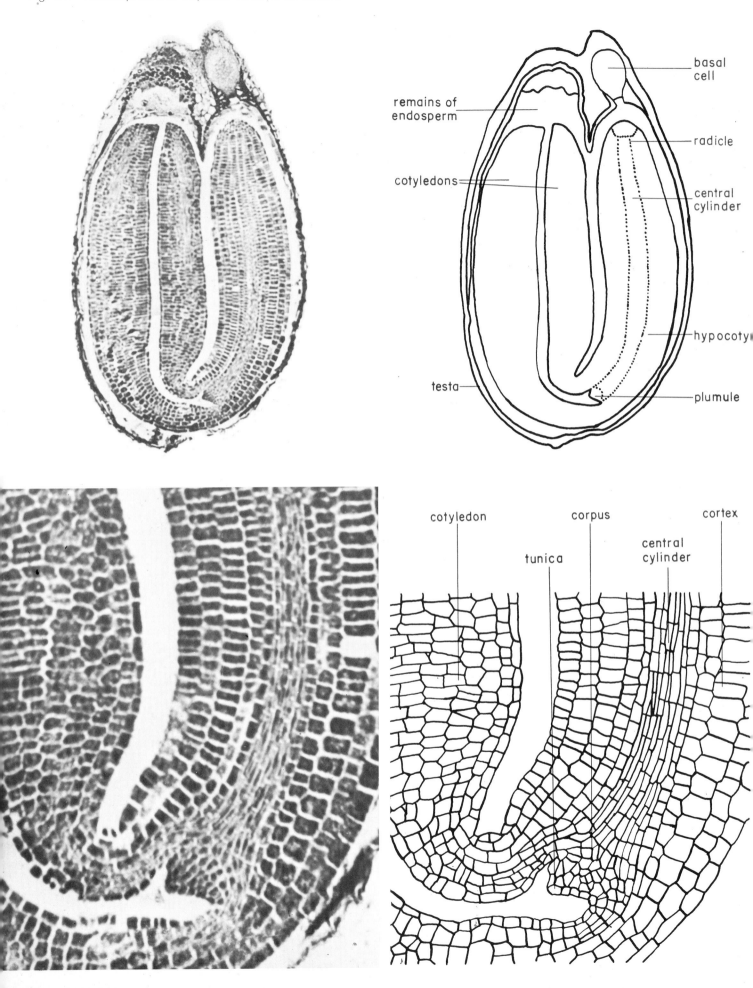

basal
cell

remains of
endosperm

cotyledons

testa

radicle

central
cylinder

hypocotyl

plumule

cotyledon   corpus   cortex

tunica   central
cylinder